THE COVERT COLONIZATION OF OUR SOLAR SYSTEM

HERBERT DORSEY

outskirtspress

DENVER, COLORADO

Outskirts Press, Inc.
http://www.outskirtspress.com

ISBN: 978-1-4787-6883-8

Outskirts Press and the "OP" logo are trademarks belonging to Outskirts Press, Inc.

PRINTED IN THE UNITED STATES OF AMERICA

Table of Contents

Introduction

This work is the result of the research of many others, which I have collected together in one place to make it easier for the reader. So, I give my grateful thanks to all of these researchers.

Dr. Michael Salla, a person I have had the good fortune to meet with an number of times at *Uncle Roberts* in Kalapana on the Big Island of Hawaii, and an author, who I consider the master of documentation, is one researcher that inspired this work. More can be learned about his very well documented and informative works at: http://exopolitics.org/

David Wilcox, who may actually be the reincarnation of Edgar Casey, was another researcher that inspired this work. He has a treasure trove of information on higher realities and secret space programs on his website: http://divinecosmos.com/

There are many more researchers that are so numerous, I cannot mention them all here, but will do so later in this book. My thanks goes out to them all as well.

Also, this work would not have been possible without the actual persons who worked inside these highly secret programs who have come forward to tell the public what is really going on in outer space. They will be revealed later in this work. My grateful thanks also goes out these courageous individuals as well.

While much of this information is in the public domain, I have added more to the subject from my own researches in the area of free energy, antigravity technology and secret societies, and from persons that actually worked on classified projects who trusted me not to reveal their identities.

Many of you - who have not had access to the many sources of information that I, fortunately, have had - may find the information in this book quite difficult to believe. That is quite understandable. The public has been fed a false history since the beginning of the twentieth century by the powers that be.It is now time for the whole truth to be revealed. This book is my continuing effort to reveal a portion of that truth.

One lie put forth by the controlled news media was that the Nazis were defeated at the end of World War II. While the general public was greatly relieved by this bit of disinformation, high level Nazis were busily infiltrating the government, military, intelligence agencies, and industry of the United States.

They had plenty of help from their fascist friends - heads of corporations, secret societies, like Skull and Bones, White Russians, Knights of Malta and Jesuits in the United States. It is true that the German army surrendered at the end of the war - but the Nazis certianlly did not!

Others may disbelieve that the Germans reconstructed a secret underground city in the Antarctic accessible via submarines that continued after the war. Or that they had created flying saucers which could operate under water, in the atmosphere or in outer space.

Rocket technology was a lesser technology the Paperclip German scientists shared with the U.S. after the war. Their anti gravity technology they largely kept to themselves - at least at first.

Many will disbelieve that the Nazis also landed on the Moon as early as 1942, using their antigravity flying disks and created a base there. Perhaps this book will provide enough evidence to open your mind, at least to the possibility.

But it gets even more interesting. Many of the same companies that were secretly financing the Nazi rise to power before World War II and after the war, assisted in their covert takeover of the constitutional government of the U.S., combined their resources to create an Interplanetary Corporate Conglomerate, beyond the control of any government, which used advanced technology to secretly colonize Mars and many moons of the solar system, using slave labor, much like the Nazis did during World War II.

In my previous writing of this genre, *Secret Science and the Secret Space Program,* much of what is in this book has already been documented.

However, I have discovered so much new information since writing that book, that a new book is called for. Here it is - *The Covert Colonizing of our Solar System.*

The First Secret Space Program

In the first part of the twentieth century, Germany was, by far, ahead of other countries in science and technology. In fact, Germany would have won World War I if the United States had not entered the war on the side of England. German U-boats had encircled Britain and created a virtual blockade of that nation because of the tremendous tonnage of shipping sunk by these submarines.

And as historians now realize, the entry of the U.S. into that war was a direct result of the Rothschilds influence in the United States as quid pro quo between Lord Rothschild and Lord Balfour for England issuing the Balfour Declaration, stating British intention to create a nation for the Jewish people in Palestine.

The Rothschilds had tremendous influence through their agents and bankers in the U.S. One notable Rothschild agent was Mandell House, a close advisor to President Wilson. House chose every cabinet position in the Wilson Administration and would later be involved with the creation of the treaty of Versailles. House advised President Wilson not to veto the Rothschild created and unconstitutional Federal Reserve Act of

1913. He also advised President Wilson to declare war against Germany.

After the Balfour Declaration, the Jewish people collectively threw their weight behind England and declared "war" against Germany even though Germany, up to then, had treated the Jewish people better than most European countries. This "war took the form of boycotts of German goods and bad press against Germany in other countries. This "war", along with the Jewish led Bolshevik revolution in Russia, would contribute to future German anti-Jewish feelings in the years following World War I.

The Bolsheviks were largely financed by the Kuhn Loeb Bank of New York and the Max Warburg Bank of Hamburg Germany. Jacob Schiff, director of Kuhn Loeb, was an agent of the Rothschilds and Max Warburg was related to the Rothschilds by marriage. The banking house of Rothschild was, before the French revolution, appointed to be the guardians of the Vatican Treasury.

So, secretly behind the Jewish leaders of the Bolsheviks was the Vatican and their Jesuit agents, like Joseph Stalin, SJ. The Jesuits historically employed prominent Jewish people to carry out their secret agendas. Then, if their plans went awry, the Jews would be blamed and not the Vatican.

The Vatican had a rival in the Eastern Orthodox Christians, which intensified after the Fourth Crusade, which was directed at the then Eastern Orthodox Christian capital of Constantinople. Later, when the Muslims invaded Constantinople, the Eastern Orthodox Christians moved their capital to Moscow. The Russian Czar was the guardian of Eastern Orthodox Christianity.

So, the Vatican backing of the Bolsheviks Revolution was part of their long range plan against the Eastern Orthodox Christians and their greater plan of eventual global domination.

After World War I, Germany was placed, by the treaty of Versailles, in an economically unfavorable situation by having to pay exorbitant war reparations payments to the countries damaged by the war. These payments would be handled by the, newly created, Bank of International Settlements (BIS) in Switzerland. Also, Germany was not allowed to rearm it's self.

These factors created a lot of resentment among the German people. Before the U.S. entry into the war, Germany, with their vastly superior U-Boats had encircled England and had created a virtual blockade of England and was winning the war. Germany had offered a cease fire with the extremely fair terms that everything would return to just the way they were before the war. Now, they suffered under the unfair terms of the Treaty of Versailles.

Next came the global depression, largely caused by the money manipulations of the central bankers. It was bad in the U.S. but even much worse in Germany which suffered around 50% unemployment. On top of all that, was rampant inflation in the Weimar government of Germany, where it took a wheelbarrow of German Marks to go shopping for groceries.

After Hitler's entry on the scene all of that changed. The Third Reich was being financed by the J. Henry Schroder Bank, with Allen Dulles as a director, in England and the Union bank of New York, whose directors included William Averell Harrriman and Prescott Bush. Allen Dulles would also be representing Germany's primary petrochemical company, I.G. Farben through his law firm, Cromwell and Sullivan which had an office in Berlin.

Before long, there was full employment in Germany and the German Mark regained it's value. Hitler became Germany's hero. But, little did the people of Germany realize the path towards destruction Hitler was leading his people.

Being extremely resourceful people, the Germans developed in other areas. One of these areas was in metaphysics. A number of societies were created by private groups interested in metaphysics, an interest developed long before the world war. Where physics deals with the world of matter and energy, metaphysics deals with the more subtle planes of existence (ie. spirit, astral, and etheric matter.)

At this time ether (or aether) was considered to be the substrate of the universe by most physicists. Aether was required to transmit waves of light. One scientist, Dr. Schapeller was experimenting with etheric matter and had discovered what he called "glowing magnetism", which was a thousand times stronger than normal magnetism.

Another scientist, Victor Schauberger was also experimenting with vortex phenomena that could only be explained using these more subtle forms of existence. So, there was some experimental proof of the metaphysics and these secret societies took the subject quite seriously.

The use of mediums to channel information from the spirit plane was also being used in these metaphysical societies. So, there is nothing "New Age" about this practice.

Maria Orsic was a famous medium in Germany. She was born in Vienna, Austria on October 10, 1895. Her father, Tomislav Orsic, was a Croatian immigrant from Zagreb, her mother, Sabine Orsic, was from Vienna. Later, Maria Orsic moved to

Munich, Germany.

In Munich, Maria was in contact with the Thule Gesellschaft and soon she created her own circle together with Traute A. from Munich and several other friends called the Alldeutsche Gesellschaft für Metaphysik (All German Society for Metaphysics). Later, the name was changed to Vril Gesellschaft (Vril Society).

The members were all young ladies. Both Maria and Traute were beautiful ladies with very long hair; Maria was blond and Traute was brown-haired. They had long ponytails, a very uncommon hairstyle at that time. This became a distinctive characteristic in all the women who joined the Vril Society, which was maintained till May 1945. They believed that their long hair acted as cosmic antenna to receive alien communication from beyond.

On February 10, 1917 Maria Orsic, fell into a trance or coma that lasted several hours, in which she claimed that beings of light from another world had communicated with her. She had reoccurring incidences of this communication in the days that followed. She was told not to tell anyone about her communication with them except the mediums Traute, Gudrun, Sigurn and Heike.

She visited her medium friends and explained her experience to them. They were quite understanding and even told Maria that they were expecting her to come to them. Later, the beings of light informed Maria that they were from the Star system they called Aldebaran (known as Alpha Tauri in the Taurus Constellation to our astronomers).

The telepathic communications with Maria Orsic contained two types of information; (1) metaphysical revelations, history of ancient civilizations on our planet and the true origins of the human race, (2) technical information on how to build a super flying machine. Also, information in an unknown script was given through trance automatic writing, which remained to be translated.

The script was given to the "Panbabylonists", a circle close to the Thule Society which was integrated by Hugo Winckler, Peter Jensen, Friedrich Delitzsch and others. It turned out that the mysterious language was actually ancient Sumerian. Sigrun, from the Vril Gesellschaft helped translate the language and decipher the strange mental images of a circular flight machine.

Maria Orsic could not understand the technical information and asked her father, Tomislav, to help her understand what it meant. He also was baffled. However, he was acquainted with Professor Winfried Otto Schumann, a scientist, who taught at the Technical University of Munich. After Mr. Orsic conveyed Maria's information to Dr. Schumann, he became quite interested.

Dr. Schumann was already familiar with Victor Shauberger's work with spiraling air and water implosion technology leading to levitation and Dr. Schapeller's work with "glowing magnetism" and extracting energy from the aether. Therefore, he thought the information relayed by the Orsics, which didn't contradict the other's work, could have potential.

The concept of "other science" (or "alternative science") matured during this time and the following years. Because of the financing difficulties it took three years until the flying machine project started taking shape.

Both the Vril and the Thule society members contributed to the effort. By 1922, parts for the machine began arriving independently from various industrial sources paid in full by Thule and Vril. A team of engineers and investors was organized to build the super flying machine which they dubbed the Jenseitsflugmaschine (JFM) (otherworld flying machine).

By 1922 development of a working prototype was underway. It consisted of a disk 8 meters in diameter, over which was a parallel disk 6.5 meters in diameter and under which was a 7 meter disk. All three disks had a 1.8 meter diameter hole in the center that enclosed 2.4 meter high propulsion and control unit.

The top and bottom disks were counter rotated and the center one was stationary as was the central propulsion and control tower.

Magnets were also added to the two counter rotating disks which were of a conductive material. Electric current could be drawn off between the inner and outer side of the disks, as homopolar generators work, powering the craft when a certain rotational threshold was reached.

Also the counter rotating disks and magnets created counter rotating torsion fields. These counter rotating torsion fields, when of sufficient strength, shield the craft from the space time continuum - effectively nullifying both gravity and apparent inertia inside this field.. Also, there seemed to be an extra dimensional effect. (1)

On March 22, 1922, the first model was tested and failed miserably. It rose about 50 feet into the air. It spun around like a giant pinwheel spouting fire and disintegrated. The pilot barely escaped with his life.

They went back to the drawing boards and Maria Orsic went

into another trance to get further guidance from the light beings. Several days later, Maria returned to Schumann giving him some new notes and pictures on how to rectify the problem.

A new twist was that Maria informed Schumann that mental control had to be used to fly the machine. Schumann was ready to throw in the towel.

Later, he continued on with the project when Maria presented him with the complete plans and instructions for a "Head Band Mental Command Device." As events progressed, Dr. Schumann not only became a father figure for Maria but also is considered the father of the German Flying Saucers.

By December 17, 1923, the second model of the Jenseitsflug-maschine (JFM2) was tested.

During this time Maria and Sigrun made 8 visits to the hangar where the craft was located and gave their channeled findings to the engineers before the test. This remote controlled test was quite successful; the craft flew at quite a high speed for 55 minutes.

However after the craft landed, it looked very weather worn instead of brand new. Maria explained this was because the craft was flying in a parallel dimension that caused alterations in the materials of the craft and could have the same effects on people flying in the craft. This explanation horrified the engineers present. Dr. Schumann was not, however, dismayed because he had Maria Orsic, who could tap into the metaphysical world to solve any problems.

It was experimented on for two years before being dismantled and stored at the Messerschmidt works at Augsburg. The financing for this project can be found in the accounts of several industrial companies, mentioned under the code JFM. It is certain that the Vril mechanism is descended from the machine for flight to the beyond but it has been indexed as the Schumann SM levitator.

In principle, the machine for flight to the beyond had to engender an extremely strong field around itself and its immediate vicinity which made all of the surrounding space, comprising that of the machine and its occupants, a microcosm completely independent from our cosmos.

At its maximum power, this field would be completely independent from all the forces and influences of our Universe, such as gravitation, electromagnetism, radiation, as well as any kind of matter. It could move at will in any gravitational field without one sensing it, or feeling forces of acceleration.

The JFM2 levitator unit was further developed by Schumann and others into RFZ (Rundflugzeug) class of flying saucers in 1934 and the Vril and Haunebu disks of 1939-1945. Dr. Schumann is considered the inventor of the Schumann Levitation Disk.

At the beginning of 1943, planning began for a spaceship in the shape of a cigar, the 139m long Andromeda craft, which was to be built at the Zeppelin works. It would transport several saucer shaped spacecraft,one Haunebu II, two Vril I and two Vril II. for interstellar long duration flights.

In late November 1924, Maria Orsic visited Rudolf Hess in his apartment in Munich, together with Rudolf von Sebottendorf, the founder of the Thule Gesellschaft.

Sebottendorf wanted to contact Dietrich Eckart, who was a

famous former Thule member that had deceased one year before. To establish contact with Eckart, Sebottendorff and other Thulists (amongst them Ernst Schulte-Strathauss) joined hands around a black-draped table.

Hess found it unnerving to watch Maria Orsic's eyeballs rolling back and showing only whites, and to see her slumping backward in her chair, mouth agape. However Sebottendorff smiled in satisfaction as the voice of Eckart started coming out of the medium. Eckart announced that he was obliged to let someone else's voice come through, with an important message.

A weird voice then identified itself as "the Sumi, dwellers of a distant world, which orbits the star Aldebaran in the constellation you call Taurus the Bull".

Hess and Schulte-Strathaus blinked at each other in surprise. According to the voice, the Sumi were an humanoid race who had briefly colonized Earth 500 million years ago. The ruins of ancient Larsa, Shurrupak and Nippur in Iraq had been built by them. Those of them who survived the great flood of Utnapishtim (the Deluge of Noah's Ark) had become the ancestors of the Aryan race.

Sebottendorff remained skeptical and asked for proof. While Maria was still in a trance, she scribbled several lines of queer-looking marks. Those marks turned out to be ancient Sumerian characters, the language of the founders of the oldest Babylonian culture.

The Vril mediums had received precise information regarding the habitable planets around the sun Aldebaran and they were willing to plan a trip there.

This project was discussed again the 22nd January 1944 in a meeting between Hitler, Himmler, Dr. W. Schumann and Kunkel of the Vril Gesellschaft. It was decided that a Vril 7 "Jäger" would be sent through a dimension channel independent of the speed of light to Aldebaran.

The solar system of Aldebaran is 68 light-years from Earth, and two inhabited planets which constitute the Realm of Sumerian orbit around their sun.

The inhabitants of this solar system are subdivided into masters, White God-like people (Aryans) and other different human races.

These latter developed because of climatic changes on the individual planets, and were the result of a degeneration of the God-like people. These mutants came to have a spiritual development inferior to the God-like people.

The more the races mixed together, the more their spiritual development was degraded. Consequently, when the sun (Aldebaran) began to expand, they were no longer able to make interplanetary voyages like their ancestors; it had become impossible for them to leave their planets.

Thus the inferior races, totally dependent on the masters, came to be evacuated in spaceships and taken to other habitable planets. In spite of the differences, there was respect between these two races, they did not encroach upon each other's living space - in contrast to Earth.

The race of the masters, the White God-like people, had started to colonize other planets similar to Earth 500 million years ago, following the expansion of the Aldebaran sun and the growing heat resulting from it, which made the planets uninhabitable. It was said that they colonized the planet Mallona (also called Maldek, Marduk, or Phaeton by the Russians) in our solar system, which existed at that time between Mars and Jupiter, where the asteroids are found today.

The Vril Society members thought that the Aldebaranians landed later on, when the Earth became slowly habitable, in Mesopotamia, and that they formed the dominant caste of the Sumerians.

These Aldebaranians were called White God-people. Moreover, the Vril telepaths received the following information: the Sumerian language was not only identical to that of the Aldebaranians, but it also had tones similar to German, and the frequency of the two languages was almost identical.

An important meeting of the Vril Society took place around Christmas 1943 at Kolberg, a seaside resort by the North Sea, which the mediums Maria and Sigrun attended.

The principal subject was the Aldebaran Enterprise. The mediums had received precise information about the inhabited planets orbiting the Aldebaran sun, and a voyage was planned to go there.

On 2 January 1944, Hitler, Himmler, Künkel and Dr Schumann (both of the Vril Society) met to talk about this Vril project. They wished to travel by means of a large spacecraft, the Vril 7, to Aldebaran, through a dimensional channel.

According to N. Ratthofer (a writer), a first test flight in the dimension channel took place in late 1944 on a Vril 7 named *Odin*. The test flight almost ended in disaster because after the flight, the Vril 7 looked "as if it had been flying for a hundred years". Its outer skin looked aged and had suffered damages in several places. This seemed to be a hazard of extra dimensional travel.

Another improved type of saucer developed from information from Maria Orsic and the Vril Society was the Vril 7 Geist which was built at Arado-Brandenburg and flew in 1944. Maria was able to fly this model telepathically without the Mind Control Head Band built earlier. Maria was spiritually evolved and did not wish for the Vril 7 Geist to be used for war. So, she stated that further improvements were needed and the saucer was sent to a hangar in Munich for further improvements.

She also had two smaller saucers of 27 foot diameter built. Dr. Schumann recruited 4 engineers to build these craft. She was secretly planning to escape Germany and join her Aldebaran friends who had already informed her that Germany would lose the War. (2)

Another metaphysical group was the SS Institute Ahnenerbe. Founded on July 1, 1935, by Heinrich Himmler, Herman Wirth, and Richard Walther Darre. The Ahnenerbe later conducted experiments and launched expeditions in an attempt to prove that Aryan Nordic populations had once ruled the world.

An expedition was mounted the under the patronage of SS Institute Ahnenerbe to Tibet in May, 1938 called *Die Erste Deutsche SS-Tibet-Expedition* headed by Dr. Ernst Schaefer, SS-Hauptsturmfuehrer. Schaffer, a German hunter and biologist, participated in two expeditions to Tibet, in 1931–1932 and 1934–1936, for sport and zoological research. The Ahnenerbe sponsored him to lead a third expedition (1938-1939) at the official invitation of the Tibetan Government.

The visit coincided with renewed Tibetan contacts with Japan. A possible explanation for the invitation is that the Tibetan Government wished to maintain cordial relations with the Japanese and their German allies as a balance against the British and Chinese. Thus, the Tibetan Government welcomed the German expedition at the 1939 New Year (Losar) celebration in Lhasa.

Schaffer stated, "The primary aim of this expedition was an holistic creation of a complete biological record of Tibet alongside a synthesis of inter-relating natural sciences with regard to geography, cartography, geology, earth magnetics, climate, plants, animals and mankind."

The expedition was formed of six men. Schaefer recruited young, fit men who would be well suited for an arduous journey. These men were; Karl Wienert, age 24, who was the geologist, Edmund Geer, also 24, was the expedition technical expert and organizer, Ernst Krause, age 38 was filmmaker and entomologist, Bruno Beger, age 26, was an expert on racial science called in German "Rassekunde" and also the student of the team's anthropologist Hans Gunther. Each man was a member of the Nazi SS.

Ernst Schäffer described his experiences during the expedition. During the festivities, he reported, the Nechung Oracle warned that although the Germans brought sweet presents and words, Tibet must be careful: Germany's leader is like a dragon.

Tsarong, the pro-Japanese former head of the Tibetan military, tried to soften the prediction. He said that the Regent had heard much more from the Oracle, but he himself was unauthorized to divulge the details. The Regent prays daily for no war between the British and the Germans, since this would have terrible consequences for Tibet as well. Both countries must understand that all good people must pray the same. During the rest of his stay in Lhasa, Schäffer met often with the Regent and had a good rapport.

On the expedition, in addition to meeting with high Tibetan monks, (the equivalent to political leaders, at the time in Tibet) Dr, Schaefer's team also discovered and met with leaders of an underground civilization called the Agarthans, who's existence was usually hidden by the high Tibetan monks, but now was revealed.

The underground realm of Agarttha was first introduced to the Western world in 1886 by the French esoteric philosopher Alexandre Saint-Yves d'Alveydre with his book *Mission de l'Inde*. Saint-Yves's book maintained that deep below the Himalayas were enormous underground cities, which were under the rule of a sovereign pontiff known as the Brahâtma. Throughout history, the "unknown superiors" cited by secret societies were believed to be emissaries from this realm who had moved underground at the onset of the Kali-Yuga, the Iron Age.

Ruled in accordance with the highest principles, the kingdom of Agarttha, sometimes known as Shambhala, represents a world that is far advanced beyond our modern culture, both technologically and spiritually. The inhabitants possess amazing skills their above ground counterparts have long since forgotten. In addition, Agarttha is home to huge libraries of books engraved in stone, enshrining the collective knowledge of humanity from its remotest origins. Saint-Yves explained that the secret world of Agarttha, and all its wisdom and wealth, would be made available for humanity when Christianity and all other known religions of the world began truly honoring their own sacred teachings.

The Agarthans, who were about six feet tall and were blue eyed blondes, looked to the explorers exactly like what they were searching for – the forbearers of the Aryan race.

At first, they claimed that they were extraterrestrials from the Pleiades star system. Their claim was given credibility by demonstrations of their flying saucer craft of their *Silver Fleet*. This falsity was part of the Agarthan strategy for dealing with surface people so they wouldn't go searching for the entrances to their underground cities.

However after a while, the German explorers gained the trust of the Agarthans who then revealed to these explorers that they lived in one of a number of underground cities on the planet.

They even mentioned cities under Antarctica and in particular one abandoned city under the ice cap in a cave that could be reached by sea using submarines that the Germans possessed. This information was quickly radioed back to their German headquarters.

On December 17, 1938 the New Swabia Expedition left Hamburg for Antarctica aboard the *MS Schwabenland* led by Captain Alfred Ritscher, a captain in the Kriegsmarine. The expedition arrived in Antarctica on January 19, 1939 at 4 degrees 30 Min. West and 69 degrees 14 Min. South off the Princess Martha coast.

The *MS Schwabenland* had a crew of 24 persons in addition to the expedition force which consisted of 33 members. The ship also had an float aircraft on board which they could catapult into the air and recover from the sea with a huge onboard crane.

This aircraft dropped 1.2 meter-long aluminum arrows, with

30 centimeter steel cones and three upper stabilizer wings embossed with swastikas, onto the ice at turning points of the flight which marked out the border of a large area from between 20 degrees East and 10 Degrees West. This area was geothermally active and contained hot springs and some ice free land. The area also enclosed the hidden abandoned underground city revealed by the Agarthans.

German U-boats were also sent into the region and before long, discovered the abandoned city under the ice cap. The city, built into a huge cave, parts of which were under sea level, was in a state of disarray.

More U-boats and men were sent and the city was reconditioned to serve as a secret German base which they named Neu Berlin. Electric power was supplied by geothermal steam generators much like in Iceland.

Also, some of the abandoned technology of the city was taken back to Germany for reverse engineering. A number of Germans were left there to live on a full time basis and Nue Berlin was regularly resupplied by submarine with food from South America and equipment from Germany. (3)

Russian UFOlogist Konstantin Ivanenko, as reported in *UFO ROUNDUP*, Volume 9, Number 3, January 21st, 2004 wrote the following:

> "If you had been a Wehrmacht soldier at the bombed-out railroad station in Poltava, a city in the Ukraine, during the summer of 1942, you may have seen a very strange-looking military unit on the march, heading for a waiting passenger train.

The unit consisted of women, all of them blond and blue-eyed, between the ages of 17 and 24, tall and slender, their sensational figures encased in striking sky-blue uniforms.

Each woman wore an Italian-style garrison cap, an A-line skirt with the hem below the knee, and a form-fitting jacket with the insignia of the SS. You might have thought the SS had recruited a platoon of high-class call girls, but the truth was far stranger than that.

You would have been looking at Reichsfuhrer-SS Heinrich Himmler's latest brainstorm—the Antarctisches Seidlungensfrauen [Antarctic Settlement Women or ASF]."

Ivanenko claims that the SS's *Rasse und Seidlungshauptamt* (Race and Settlement Bureau or RuSHA) was the agency responsible for selecting women for the *Antarctisches Seidlungensfrauen*.

He also claimed that as many as 10,000 Ukrainians with acceptable racial purity (out of the more than half a million Ukrainians deported during the war) were transported, not to munitions factories in Germany, but to the German Antarctic! Now, the Germans had both slave labor to reconstruct Neu Berlin and racially pure Aryan women to multiply the population of Neu Berlin in the future.

Famous researchers of the Third Reich's Antarctic mysteries, Renato Vesco, Vladimir. Terziysky, and David Hatcher Childress, claim that from 1942, thousands of concentration camp prisoners, prominent scientists, pilots, politicians with their families, and members of the Hitlerjugend were taken to the South Pole via submarine. Some scientists believe that a German base still remains in the Antarctic. Moreover, it is even said that there is an underground Aryan city called Neu Berlin, which has a population of two million people.

Not only had the Agarthans shown the Germans where this abandoned underground city was located, the Agarthans had also agreed to a treaty with the Third Reich which allowed them to rebuild this abandoned city. The Agarthans also had two other underground cities in Antarctica, further inland, that they still occupied.

The Germans of the Third Reich also created other underground bases in Tierra Del Fuego and other portions of South America.

So, now the Germans had two different sources of information on the means to build flying saucers – Maria Orsic channeling the Aldebarians and the Agarthans. But, this was not all. They also had sent spies into the United States to recover work that Nikola Tesla had been working on. One such spy was George Sherif, a trusted assistant of Tesla's. One unpatented project of Tesla was a flying vehicle, which operated on electromagnetic – not aerodynamic principles.

Also, the Germans were familiar with the work of Thomas Townsend Brown who had experimentally discovered the connection between electricity and gravity in the 1920s. And, the Germans had access to the work of Victor Schauberger who had developed the saucer shaped, levity disks. In addition the Germans were studying ancient Hindu scriptures which described ancient flying Vimanas which used rotating mercury to generate flight.

So, a number of different German flying saucer development programs were going simultaneously before and during World War II. The secret V-7 projects were in fact a series of circular aircraft research designs developing radical new weapons platforms that could rise and descend vertically.

They also could fly at high speeds and very high altitudes using a new "implosion" motor design and other electro-gravitic energy conversion principles producing a self-sufficient electro-statically generated force for power.

Such engines were manufactured by Allgemeine Elektrizitäts Gesellschaft (AEG), the great electrical giant of industrial Germany, for this program as early as 1944. These power devices generated their own electro-gravity field which neutralized the Earth's natural gravity, in effect making them weightless. Another new engine principle used helium for fuel. This concept has been rediscovered in the Joseph Papp noble gas engine.

There were at least three major developments within the V-7 program. There were the Haunebu models 1 through 3, at least; and Vril models 1 through 7 as well, and most likely beyond that. Then there was the cylindrical mother-ship carrier project designed to transport the Haunebu and Vril craft in flight. The cylindrical carrier craft were over 100' in diameter and could carry one or more of the Haunebu models and 3 or more of the smaller Vril models, all internally, and could launch and recover them in flight. Research along these lines had been carried out since early in 1941.

Since 1935 Thule had been scouting for a remote, inconspicuous, underdeveloped testing ground for their Thule designed craft. Thule found a location in Northwest Germany that was known as Hauneburg.

At the establishment of this testing ground and facilities the SS E-IV unit simply referred to the new Thule disk as Haunebu in

1939 and briefly designated it the RFZ-5 along with Vril's other machines which included the Rundflugzeug RFZ-1 through the RFZ-4.

At a much later time in the war, as production of these craft was to commence, the Hauneburg site was abandoned in favor of the more suitable Vril Arado Brandenburg aircraft testing grounds. Although designated as part of the RFZ series the Haunebu disk was actually a separate Thule product constructed with the help of the SS E-IV unit while the RFZ series were primarily built at Arado Brandenburg under Vril direction up to the RFZ-4 disc. Three models were built, the Haunebu I, Haunebu II and Haunebu III.

The early Haunebu I craft of which two prototypes were constructed were 25 meters in diameter, had a crew of eight and could achieve the incredible initial velocity of 4,800 km/h, but at low altitude. Further enhancement enabled the machine to reach 17,000 km/h! Flight endurance was 18 hours. To resist the incredible temperatures of these velocities a special armor called Victalen was pioneered by SS metallurgists specifically for both the Haunebu and Vril series of disk craft. The Haunebu I first flew in 1939 and both prototypes made 52 test flights.

In 1942, the enlarged Haunebu II of 26 meters diameter was ready for flight testing. This disk had a crew of nine and could also achieve supersonic flight of between 6,000-21,000 km/h with a flight endurance of 55 hours.

Both it and the further developed 32 meter diameter Do-Stra had heat shielding of two hulls of Victalen. Seven of these craft were constructed and tested between 1943-44. The craft made 106 test flights.

By 1944, the perfected war model, the Haunebu II Do-Stra (Dornier STR Atosphären Flugzeug) was tested. Two prototypes were built. These massive machines, several stories tall, were crewed by 20 men. They were also capable of hypersonic speed beyond 21,000 km/h. The close of the war, however, prevented Dornier from building any production models.

The Haunebu III was a larger 71 meter diameter. A lone prototype was constructed before the close of the war. It was crewed by 32 and could achieve speeds of between 7,000 - 40,000 km/h! It had a triple Victalen hull. It is said to have had a flight endurance of between 7-8 weeks! The craft made 19 test flights and was likely used to evacuate the Thule and Vril people along with Hans Kammler at the end of the war. Both societies disappeared at the close of the war.

According to researcher Vladimir Terezski and whistleblower Corey Goode, the Germans had flown these large saucers to the far side of the Moon in 1942 and started working on an underground base there.

The first man on the moon was Kapitan Leutnant Werner Theisenberg of the Kriegsmarine. Actually, the German Navy did most of the work on the German space program, not the Luftwaffe. Ever since their first day of landing on the Moon, the Germans started boring and tunneling under the surface and by the end of the war there was a small Nazi research base on the Moon.

This Moon base construction was only made possible by a treaty agreed to between the German government and the Draco federation near the end of the war. The Moon is occupied by a number of different extraterrestrial groups and is used as a diplomatic meeting ground between these groups. The Draco federation is one such group and they granted permission to the Germans to build a base there.

Actually, during the last years of World War II, the informed people of Germany knew that they would lose the war, in spite of all the propaganda to the contrary.

Much of the strategy then was to relocate as much of the advanced German technology to other locations like Argentina, Antarctica and the Moon. Hans Kammler was in charge of this project and succeeded in moving much of this material before the allies over took the areas where they were located.

A parallel project was to relocate as much gold and other financial resources to neutral countries like Switzerland, Spain and Argentina. This was headed by Borman and Hjalmar Schacht and then largely carried out by Otto Skorzeny after the war. Both projects were designed to create the secret Nazi International that would succed the Third Reich. (4)

As the Allied Command pondered German intentions in Spain, it was aware that although Spain was neutral, General Franco, because of German threats, was under the German thumb.

London, therefore, concluded that Spain's importance to the enemy lay in the use of her Atlantic ports. Reports arrived regarding a new fleet of giant German submarines approximately 400 feet long and several decks high. Agents reported sightings of these subs in the vicinity of Aymonte and Hulva, Spain. and also at Baltic and Norwegian ports.

In 1944-1945 it was confirmed that the loading of these subs at Baltic ports with unusual machinery and equipment was secretly being carried on. The Norwegian underground picked up the super subs' trail. These reports pieced together told a tale. The giant underwater megaliths had left Germany, thence to Norway and along its coast northward to avoid Allied shipping lanes, and then west from Narvik toward Iceland in the North Atlantic. From a point below Iceland the subs steered an oblique southerly course which eventually took them to the Atlantic ports of Hulva and Aymonte, Spain.

Involved in this drama was one of the giant German submarines. She was one of those built secretly in 1944 and carried a cargo of top secret German plans, documents and prototypes of new inventions. The sub was in the North Atlantic at an approximate position of 14 degrees West and 3 degrees North, when her oxygen supply gave out due to malfunction of equipment. Unable to stay submerged, the leviathan slowly ascended from a depth of 2000 feet and its 12" thick hull of steel broke surface of the cold Atlantic at midnight on 23rd of April, 1945, within a mile of two British cruisers.

Up went flares as the British ships opened fire on the German sub. Eight inch shells straddled the huge craft to get range, as an odd-shaped German gun appeared on the sub's hull. A pencil beam of laser horned in on the cruiser *Camden*.

There was no battle thunder or fury from the sub. The ray silently pierced the darkness and flares, and in seconds a 20 foot diameter hole was cut from port to starboard through the first surface ship.

Like a toy boat suddenly filled with water, the cruiser sank horizontally with a hissing of steam from the white hot steel hull.

Then the beam moved onto the second cruiser, *Hanover*, and as another 20 foot hole was opened, she burst into flames, and settled down in less than 30 seconds.

Unknown to the allies in late 1944, the secretive and orderly German evacuation was proceeding well. Her top personnel which were needed to continue the Third Reich elsewhere were being removed by round wing planes and the super subs, the chief vehicles used among several withdrawal methods.

But, Germans like Von Runstedt, from whose area the German flying saucers were evacuating key personnel, refused to give travel priority to those Nazis responsible for Jewish exterminations.

The truth is that most of the regular German general staff had disdain for these Nazis who were not wanted in the new Germany destined to rise in another part of the world. Furthermore, German embarkation officers in the Hulva and Aymonte also refused to place key Nazi killers on board the super subs.

Major Otto Skorzeny, Hitler's tough deputy, had gone personally to Von Runstedt in December, 1945 and demanded seats for himself and his top henchmen on the departing flying saucers. Von Runstedt had refused and so had Von Schusnick, the chief pilot. Thus the Nazis had to find their own way out of Europe to escape allied vengeance in the coming Nuremberg War Trials.

The German people worshiped their Furher, Hitler. And Hitler was a true believer in a a new utopia for Germany. Therefore, Hitler was the key to the German evacuation. Hitler and Eva Braun were legally married on April 29, 1945 but their first born son Adolph Hitler II, was born in 1940, five years before their marriage. Hitler was said to be the father.

Back in October, 1944, a select German group, working from

the schedule compiled on August 10 in Saltzberg, decided to implement operation "Get Lost." Hitler was to be the catalyst.

All the art treasures, scientific developments, and treasury bullion which Germany possessed were first scheduled to be hidden or removed. First to be safely removed, however, would be the Fuehrer. Over Hitler's protests he was asked to pack immediately his personal possessions and leave Germany for the new land. A double stood by to assume the Fuehrer's role and he would continue under the tutelage and surveillance of Bormann, Goebbels and Ley.

Eventually, Hitler and family made it to Ayamonte, Spain. On November 7, Hitler and his family were taken on board a super sub, along with 500 other Germans. As the super sub slipped beneath the water she headed southwest.

For the next 18 days, in an 8 x 10 cabin, Hitler and his family shared living and sleeping quarters. Two leather covered chairs, four bunks and a radio for the Fuehrer and his family were the accessories." There were two doctors in attendance on board the submarine for the 500 passengers, submarine crew and Hitler and his family.

Back in New York, more OSS reports told of additional German arrivals in South America. The Germans were flooding into Belem and other river ports, as well as air strips in Brazil's Amazon Valley, Leticia in Colombia and Georgetown, British Guiana.

The Germans always appeared to be in transit. At that point OSS General Donovan personally went to Brazil to direct operations. American agents posed as rubber, precious metals and timber buyers along the Amazon and Orinoco Rivers.

In 1947, a U.S. military expedition of an aircraft carrier, a submarine, 20 surface ships and 4,000 elite Navy troops, disguised as a scientific exploration, was launched against this base under the code name "Operation High Jump".

Operation High Jump, officially titled *The United States Navy Antarctic Developments Program, 1946–1947*, was a U.S. Navy operation organized by Rear Admiral Richard E. Byrd Jr., USN (Ret), Officer in Charge, Task Force 68, and led by Rear Admiral Richard H. Cruzen, USN, Commanding Officer, Task Force 68.

Operation High Jump commenced 26 August 1946 and ended in late February 1947

The early termination of the expedition was caused by the military losses incurred by this South Polar fleet. On Jan 17, 1947, two days after Byrd's central force made the Bay of Wales, strange lights were spotted in the distance in the early morning darkness, by radioman John P. Szelwech aboard the *USS Brownson*. They did not show up on radar. This was entered into the ship's log.

About 3 hours later, the lights reappeared and rapidly approached the destroyer. Commander Gimber ordered the ship's 40 mm Bofors anti aircraft guns and 20 mm Oerlikon cannons to open fire on the rapidly approaching craft, which rapidly flew at about 200 feet above the *USS Brownson*. This anti aircraft fire seemed to have no effect on these strange craft.

This encounter was the beginning of a number of brief but fierce squirmishes that lasted over the next several weeks, in which dozens of officers and men were killed. This led to the evac-·uation of Byrd's central command from the Bay of Wales on 22 by the icebreaker *Burton Island*.

As a parting shot, the Germans attacked the retreating fleet again on February 26. This attack was witnessed by flying boat pilot, John Sayerson, who said the following:

> "The thing shot up out of the water at a tremendous velocity as if pursued by the devil and flew between the masts [of the ship] at such a high speed that the radio antenna oscillated back and forth in it's turbulence. An aircraft from the Currituck that took off just a few moments later was struck by some kind of an unknown ray from the object, and almost instantly crashed into the sea near our vessel.
>
> I could hardly believe what I saw. The thing flew without making any sound as it passed close over our ships and harmlessly through their lethal anti aircraft fire.
>
> About 10 miles away, the torpedo boat Maddox burst into flames and began to sink. Despite the danger, rescue boats went to her aid before she sank 20 minutes later.
>
> Having personally witnessed this attack by the object that flew out of the sea, all I can say was that it was frightening."

Officially, Operation High Jump's primary mission was to establish the Antarctic research base Little America IV. After eight weeks, Operation High Jump was defeated by German flying saucers and Byrd beat a hasty retreat back to the U.S.

Admiral Byrd made the following statement to a reporter for the Chilean newspaper *Brisant* while aboard the command ship, *Mount Olympus* on the voyage home:

"It was necessary for the USA to take defensive actions against enemy air fighters which come from the polar regions...Fighters that are able to fly from one pole to the other with incredible speed."

Also, Admiral Byrd, in an interview with Lee Van Atta of International news Service aboard the expeditions command ship, discussed the lessons learned from the operation. The interview appeared in the Wednesday, March 5, 1947 edition of the Chilean newspaper El Mercurio and read in part as follows:

Admiral Richard E. Byrd warned today that the United States should adopt measures of protection against the possibility of an invasion of the country by hostile planes coming from the polar regions.

The Admiral explained that he was not trying to scare anyone, but the cruel reality is that in case of a new war, the United States could be attacked by planes flying over one or both poles. This statement was made as part of a recapitulation of his own polar experience, in an exclusive interview with International News Service.

Talking about the recently completed expedition, Byrd said that the most important result of his observations and discoveries is the potential effect that they have in relation to the security of the United States. The fantastic speed with which the world is shrinking – recalled the Admiral – is one of the most important lessons learned during his recent Antarctic exploration.

I have to warn my compatriots that the time has ended when we were able to take refuge in our isolation and rely on the certainty that the distances, the oceans, and the poles were a ee of safety.

A 2006, Russian documentary made public for the first time a 1947 secret Soviet intelligence report commissioned by Joseph Stalin of the Operation Highjump mission to Antarctica. The intelligence report, gathered from Soviet spies embedded in the US, revealed that the US Navy had sent the military expedition to find and destroy a hidden Nazi base. On the way, they encountered a mysterious UFO force that attacked the military expedition destroying several ships and a significant number of planes. Indeed, Operation Highjump had suffered "many casualties" as stated in initial press reports from Chile. Here is the documentary:

https://www.youtube.com/watch?v=MwUpPwyyvLw

Much of Operation High Jump is still classified today.

At the wars end in Germany, any remaining German flying saucers were destroyed and their engineers and scientists killed by the SS before the Allies could overrun their bases and gain their technology.

However, the Allies, the U.S., Russia, Britain and Canada did get some of the technical data, partially destroyed hardware and engineers and scientists of the flying saucer program and are keeping it all top secret.

The U.S. received the lion's share of the information. Many of the Germans who worked on these projects, including Dr. Schumann, the father of the German flying saucers, were brought to the U.S. in "Operation Paperclip." All in all about 5,000 Paperclip Nazis were brought in. Later, they were hired by the military and defense contractors in their own "black projects."

So, the Hanabu saucers, which were able to operate under-water, in the atmosphere and outer space, were able to defeat the Operation High Jump forces with their superior technology which they were still progressing even further at both Neu Berlin and at underground bases in Argentina. By this time the scientists at Nue Berlin, had also developed working nuclear weapons of great power.

Also, the Germans in Nue Berlin had also formed an alliance with the Draconian Federation – a particularly nasty group of reptilians - in exchange for technology. Part of their treaty with the Draconians allowed them to construct a base on the Moon which was already partitioned between many extraterrestrial groups as a sort of diplomatic meeting ground.

This account of the German Moon base is given more credibility from testimony given by men who have actually been there, notably Randy Cramer and Corey Goode.

In particular, Corey Goode disclosed what he says he read in classified historical documents he had access to during his 20 year tour-of-duty (1987-2007) with "Solar Warden" and other secret space programs.

In his latest interview on the Cosmic Disclosure Gaiam TV series, Goode elaborated upon his earlier revelations that he was given access to "smart glass pads," similar to iPads, where he read classified information about the origins of secret space programs.

Previously he said that he was given this level of access to classified information since his duties involved familiarizing himself with many aspects of secret space program history and activities, in order to better fulfill his responsibilities as an "Intuitive Empath."

He recalls that the "smart glass pads" revealed that channelers, associated with the Vril Society, made contact with a group of Draconian extraterrestrials claiming to be humans from the Aldebaran star system.

Later, during expeditions to the Himalayas, members of the Vril Society and Nazi SS made contact with a group of Nordic-looking humans, who were part of the "Agartha Network". Goode says that the Agartha network had advanced spacecraft, named the "Silver Fleet." The Agarthans held similar beliefs to the Nazis in terms of racial purity and separate development models for humans.

This is not the first time Goode has discussed the development of secret space programs in Germany during the 1920s and 1930s under the guidance of secret societies and later the Nazi SS. In a May 20 email interview, Goode went into some detail about the role of German secret societies and the Nazi SS in developing parallel secret space programs.

Goode's stunning revelations have been investigated closely in the newly released and extremely well documented book by Michael Salla, *Insiders Reveal Secret Space Programs and Extraterrestrial Alliances*. Findings from available historical documents corroborate many of Goode's key claims regarding the role of German secret societies and the Nazi SS in developing, not one, but two secret space programs! Today, this information explains many puzzling political decisions and military actions.

The policy implications of Goode's disclosures illuminate current events in an astounding way. Not only did German secret societies and the Nazi SS cooperate in their respective space programs, but they collaborated in a joint effort after World War II to successfully infiltrate the infant U.S. secret space program.

Goode's claims that Nazi flying saucers defeated Admiral Byrd's 1947/48 Naval expedition and did overflights of major U.S. cities in the 1950's is supported by historical documents, and other whistleblowers discussed in *Insiders Reveal Secret Space Programs and Extraterrestrial Alliances*.

Among them is Clark McClelland who worked as a NASA contractor and employee for 34 years. McClelland once observed a 9 foot tall E.T. in a space suit communicating with 2 American astronauts in the cargo bay of the space shuttle, with an E.T. space craft nearby. He has an informative website here: http://www.stargate-chronicles.com

These documents and independent testimonies help respond to critical readers seeking evidence to substantiate Goode's claims, and help to gain an understanding of why this information illuminates many truths today withheld by those in back of the secret space programs.

It is clear that if Goode is even partially correct in his incredible claims, then history will have to be radically rewritten. This will be deeply disturbing to those believing that Nazism, and all it stood for, was resoundingly defeated at the end of World War II.

The Nazi International

As stated earlier, The Nazis never surrendered at the end of world war II, The Commander of the German military forces, Karl Dönitz did – not the Nazis! By the time of the end of the war, the Nazis were already working on "plan B." And, surprisingly, they had plenty of help from within the U.S.!

The definition of fascism is the joining together of government and corporations. Before the war, the U.S. was in the throes of a depression and many political ideologies ran rampant in the U.S. To some, communism held the solution to their economic suffering. At the time communism was an acceptable concept within the U.S. Communists were taking over labor unions with the slogan "Workers of the world unite".

Fascism, particularly among the industrialists was also in favor. Companies like International Telegraph and Telephone (ITT), Ford, Dupont, Standard Oil, International Business Machine(IBM), Union Bank, and others were all definitely pro fascist. They pointed to the "economic miracles" accomplished by Benito Mussolini and Adolf Hitler in Italy and Germany while the U.S. was still suffering in the depression.

Historians now realize that both communism and fascism and their respective rise to power was being financed by illuminati bankers that were financing both sides, as the Rothschilds historically were known to do.

At one point, some of these industrialists even planned a military coup against the Roosevelt Administration while charging that Roosevelt was a communists because of his "New Deal" reforms. They recruited General Smedley Butler, a World War I, hero to lead this military coup.

Butler went along with their plans at first, to discover more about who the plotters were. After learning as much as he could, he informed Congress of their plans and the coup failed.

However, none of the plotters were punished, as they were "too big to fail". Congress and the press swept the whole affair under the carpet and the event was soon forgotten.

Many of these corporations were eager to hire the Nazi scientists brought over from Germany in Operation Paperclip after the war. And many of the employers and employees still thought that that fascism was a good idea - even while mainstream America was horrified by what they had discovered of the Third Reich.

At this time period after the war, the Nazi International was being formed by men like Otto Skorzeny, Hjalmar Schacht, and Martin Bormann, who were still loyal to Hitler. Hitler and Eva Braun, after Borman had faked their suicide in the "Hitler Bunker" in Berlin, had both been smuggled via U-Boat into Argentina and were living comfortably at their Alpine like, resort at San Carlos de Bariloche in their Inalco estate. (6)

The Argentine leader, Peron was favorable to the Nazis because Briton had tried to make claims to the land south of the latitude of the Falkland Islands in both Argentina and Chile, during the first world War. Also, both Chile and Argentina had large German colonies.

This secret Nazi International, which was well documented in *The Nazi International* by Joseph P. Farrell, and by other researchers, would infiltrate U.S. and other companies around the globe, after the U.S. occupation of Germany ended.

Otto Skorzeny funded the Nazi International. And, at the same time he was employed by the CIA after the war! He assisted Martin Bormann in recovering Nazi loot deposited in neutral countries for this organization.

The original idea for this postwar organization came, not from Skorzeny, but Martin Bormann, Hitler's private secratery. At a secret meeting at the Hotel Rotes Haus, in Strasbourg on August 10, 1944, Bormann met with key German industrialists. They realized that Germany would lose the war and developed plans for a post war world.

They would immediately start transferring money, machine tools, specialty steel and blueprints for advanced technology abroad to neutral countries, to be used after the war. Also, the industrialists were encouraged to set up firms in these neutral countries.

Skorzeny was Hjalmar Schacht's nephew by marriage and both men were involved in setting up the Nazi International. The financier of the Third Reich, Hjalmar Schacht, advised the industrialists how to camouflage their assets on paper. Otto Skorzeny would handle the actual physical transfer to this wealth. Both men knew the secrets of the Nazi treasure: where it was located; how it got there; who controlled it; and the purpose it was intended for.

Also, Otto Skorzeny was the most important man in organizing *Organization der Entlassene SS Angehorige,* (ODESSA - organization for the release of former SS members).

ODESSA would become an important part of the Nazi International and assist Nazi war criminals escape justice and gain the early release from prison of ones found guilty and imprisoned.(7)

While Otto Skorzeny was in prison in Dachau, awaiting war crimes trials , money that Martin Bormann had transferred to Argentina was originally placed under the control of a trusted fellow fascist, a colonial in the Argentine military, Juan Peron.

Juan Peron had organized a military coup in Argentina and placed General Pedro Ramirez in as the new president of Argentina. But Peron was the real power behind the throne, who would later become President of Argentina. Peron had also helped the Germans set up a number of spy posts during the War in supposedly neutral Argentina.

The money, which was brought in U-boats, was placed in the Banco Aleman in Buenos Aires, by Rudolf Ludwig Freude, a German Argentinian banker who had helped Hitler many times, and who was assisted by Heinrich Dorge , a personal friend of Hjalmar Schacht.

Evita Duarte was Peron's mistress and after they were married in 1945 she became quite interested in the Bormann treasure. She persuaded Freude and Dorge that the money would be safer in an account in her name, because of the risk of the Allies confiscating German funds after the war. The both knew that Peron was the real power in Argentina and that she might be correct about confiscation and so agreed.

After the War, Bormann had disappeared but had been tried in absentia at Nuremberg and found guilty of war crimes. When Bormann failed to show up in Argentina after the War, the Perons started thinking that the Borman funds were their own property. Actually, Bormann was in Argentina. He was brought in by U-boat from Spain and hiding at San Carlos de Barluchi, but didn't dare show his face in public, for fear of arrest.

Evita was a tough and greedy woman who was the brains behind the President of Argentina, Juan Peron. But she was also famous in her own right because of her movie actress career. For public relations reasons she also had set up a number of charity organizations in Argentina. She also deposited much of the Borman funds into numbered Swiss accounts.

Later, after being found innocent of war crimes, Otto Skorzeny would set up a base in Neutral but fascist Spain, as many other former Nazi SS fled there after the war and had established a German colony there. But, he also would make trips to Argentina, where his primary mission was to recover the Borman funds.

In 1949, The Perons received Skorzeny warmly and they became friends. Skorzeny never mentioned the Borman funds. What Skorzeny did discuss was how he could help maintain order in Argentina which at the time suffered a certain amount of unrest. Skorzeny trained the Buenos Aires police in Nazi torture and Gestapo interrigation tactics. Peron was so delighted with the results that he asked Skorzeny to train all of Argentina's police forces.

While in Buenos Aires, Skorzeny discovered an assassination plot against Evita Peron from his German spies there. The plotters planned to kill her while she was visiting one of her charities. Skorzeny had his men go to the site and arrest the assassins and hold them there.

Then, he found Evita and warned her of the plot. She didn't believe Skorzeny, but invited him to come along on as protection, just in case, on her planned visit.

When their car approached the charity Skorzeny ordered the driver to stop. He then ran inside with gun drawn and soon returned with the two assassins at gunpoint. He immediately became Evita's hero. Not much later, Skorzeny would have a love affair with Evita.

She would later die of cancer. After her death, Juan Peron's popularity went down hill, as it was her PR programs that helped him win the presidency in the first place.

Peron made a lot of enemies in Argentina that wanted to kill him and was eventually ousted from the presidency in 1955 and escaped to Spain with considerable help from Skorzeny and his men.

The price? The remainder of the Borman funds. These funds were used to fininance both the Nazi International and ODESSA.

Another important man of the Nazi International was Knight of Malta, General Reinhardt Gehlen, head of the German intelligence on Eastern Europe. He had surrendered to the U.S. Army and soon was whisked off to Camp David for interviews with the OSS and Military Intelligence.

After Allen Dulles became Director of the CIA, he hired Gehlen and his entire intelligence organization and allowed them to continue to operate in Germany because the CIA possessed virtually zero intelligence in Eastern Europe and Russia. At this point, the CIA had about 50% former Nazi SS personell on it's payroll!

Interestingly, the Gehlen Organization was given complete independence of operation by the CIA. Soon, the Gehlen Organization was feeding false intelligence to the CIA, stating that Russia was preparing to invade Western Europe.

In actuality the Soviets were too busy rebuilding their war torn countries to even consider such an adventure. But this disinformation helped to fuel the cold war! So, attention was focused on Russia as a defeated Germany was soon forgotten about.

Part of this strange state of affairs is explained in my *Secret History of the New World Order*. The men that created the CIA, Allen Dulles and William Dnovan, were Knight of Malta agents of the Vatican – the same institution that brought Europe's fascists into power in the first place and helped the Nazi war criminals escape justice after World War II. Gehlen was also a Knight of Malta. The Vatican also played a large role in assisting ODESSA relocate Nazi war criminals to other countries.

In any case, the goal of the Nazi International after the defeat of Germany was not a resurrection of the Third Reich that existed under Hitler. It was to gain access to the tremendous industrial productivity of the United States – that prodigious productivity that led to the defeat of the Third Reich.

They had the advanced science but lacked the production facilities to properly construct their space faring fleet needed to colonize the solar system. They also desired this access to re-build a destroyed Germany.

The Nazi International was gradually infiltrating U.S. industry, Intelligence agencies, politics and the official U.S. space program.

Most significantly, they were infiltrating the Military Industrial Complex with their Operation Paperclip Nazis, many of whom continued with their Nazi beliefs and goals while working in American industry.

After Operation HighJump and it's failure to wipe out the Nue Berlin base, The Truman government was considering attacking the secret base with nuclear weapons. Word of this plan got back to the Paperclip Nazis in the U.S., who relayed this information to their fellow Nazis in Argentina and Antarctica.

In response, the Nue Berlin base sent a squadron of their flying saucers in 2 waves of massive overflights of the U.S. capital in Washington D.C. in 1952. This show of superior force was meant to let Washington know how easily their own city could be also be destroyed.

The Nazi International had learned that the subject of extraterrestrials had been highly classified by the Truman Administration after several crashed extraterrestrials saucers had been recovered and sent to Wright Patterson Air Field to be back engineered. The U.S. policy was to debunk the whole subject of extraterrestrials to the public.

The technology was so advanced that the U.S. had to use the Paperclip Nazi scientists to do the back engineering. It was they, that revealed this information to their comrades in Antarctica and Argentina.

So, another motive in the massive overflight of Washington D.C. by Nazi saucers was to threaten the secrecy of the extraterrestrial subject and exotic antigravity and free energy technology. If the public got wind of it's existence the energy companies would soon be out of business.

Soon, secret negotiations were opened up between the secret Nazi International and the U.S. Government using some of the Paperclip Nazis as go betweens. The Nazi International made some demands that Truman didn't like but he did agree to leave the Nue Berlin base in peace.

These negotiations continued into the Eisenhower Administration when a final secret treaty was agreed to between the two parties. Eisenhower's advisors persuaded him to agree to a treaty to bide time to infiltrate and take over the Nazi International while developing U.S. technology and bring it up to speed with the Nazis' technology to better defend the U.S.

During the same time period the Eisenhower Administration also entered into the secret Greada treaty with extraterrestrials from the Zeta Reticuli star system, for similar reasons – to bide time while acquiring superior technology to match that of the Zeta Reticulans to better be able to defend ourselves from a possible extraterrestrial attack on Earth.

But, the Nazi International was, as we have seen, was already infiltrating the U.S. industrial base with lots of help from CEOs with a fascist philosophy. So unfortunately, the Nazis won out in the infiltration race. And this was most likely attributed to the nature of corporate short term greed, while ignoring the long range good of all.

The Nazi International, once they had a firm grip on the infrastructure of the United States, the massive build out started to begin in space and the secret space programs. And they started working together with American cabal groups that had pretty much been working together the whole time, during World War I and World War II anyway.

On Earth some corporations hired their own mercenary army – often from former, well trained, U.S. military members, to guard some of their operations. For example the oil companies would use these mercenaries to guard their oil facilities in hostile countries.

In Latin America and Africa, local native tribes were opposed to drilling on their lands and started sabotaging the oil company infrastructure. These mercenaries would then have to be hired and be used to protect these facilities.

I personally knew one of these ex mercenaries who guarded oil company infrastructure in Latin America and who shot a few natives in the course of his job. He was not too proud of his work and finally quit. He tried to hide his emotional guilt in alcoholism.

In like manner, mercenaries would also be hired by defense contractors, like Lockheed Martin and Northrop Grumman and used to guard "black project" facilities of the secret space program operated by these corporations:

A one star Air Force General who worked at Wright Patterson AFB and who doesn't wish to be named revealed that even he was scared at what he saw.

These hangers aren't controlled by the U.S. Government, even though they are within U.S. property and grounds, and these programs that none of the Air Force brass were allowed access to were all black budget, covert, external groups not related to the U.S. Government or Military.

He even said the hangers were guarded by men with security like camouflage on with no insignia, no rank, no affiliation with the Air Force.

These guards were part of a completely hidden black budget program that was ultimately commissioned somehow and someway by the U.S. government, and somehow funded by black budget funds, but completely seperate from the U.S. government as an entity. Meaning, whatever happened to this group, our government was not responsible.

It was kind of like ultra clandestine CIA operations, or delta force operations where the U.S. takes no responsibility. He went into further detail stating this was even crazier than that.

These people didn't even have any names and basically didn't exists, at all. He said he would be surprised if they even used their real names when speaking to non group officials.

These guys did not exist. Not even to himself, a one star general in the AF, or his 4 star commanding generals of the AF at Wright Patterson AFB.

He also admitted that there was more than just Apollo 18, 19, 20, the Russians have had men on the moon, and we have been there back and forth more times than you can even imagine. He also stated we have bases on the moon with 100% certainty.

He also said, "Pretty soon the world will be able to commercially travel to mars and the moons of mars, we can, and have already done it. And when commercial space flight brings us to Phobos, people are in for a huge surprise when they find this.

As a matter of fact, there is an obelisk man made structure on one of the moons of Mars, Phobos, and something that no one has told you about that is the Russians, Chinese, and Americans created that in a collaborative mission."

He continued ,"we have been doing a lot of things behind the scenes in space, don't listen to Obama and the news about our lack of space travel, and our closing of our only space station **with a laugh**, but more importantly it is compartmentalized and I only heard stories from other officers about the really classified stuff, and I experienced only some of it with my own eyes...I have seen a lot of joint NASA/USAF operations documents regarding moon projects, mars projects, and space travel, ie: hidden space program during my career, everything that is in regards to national defense, I had my hands and eyes in at my command 1 star flag level position.

But, unfortunately I cannot sit here and tell you I have seen aliens, and I have proof, because even a retired 1 star general like myself has never had any access to that type of information.

I can tell you with 120% certainty that I have seen planes in triangle and full circular shapes lift off the ground, using gravity propulsion and I have seen planes speed in excess of Mach 14 and still make maneuvers at that speed, but I cannot tell you I have seen a little green man.

There are a lot of things the government doesn't want you to know, and I was and still am a part of that cover-up"

So here we see one layer of secrecy hiding another layer. Not even the Generals at Wright Patterson AFB fully know what is going on.

Similar problems are also likely at Area 51 and S-4. in the Nellis Test Range. Interestingly though, this General, like Corey Goode, predicted that pretty soon the world will be able to commercially travel into space.

Since the Russians, Chinese and the U.S are already collaborating in this secret space program, obviously the secrecy is only to keep the people of the world in the dark.

So, the Nazi International operated much like a breakaway society. The most important part of a breakaway society is how it will finance itself. It takes money, manpower and patience to create a breakaway society within an already existing civilization.

It already has been established that there were German-American banking and corporate houses in existence before World Wars I and II in which business seemed more important than national loyalty. These organizations operated secretly during both wars and afterwards.

The questions we need to ask are the following:

Why does the U.S. government allow "black budget" off the books financing of some of it's projects. Does this not bypass one constitutional function of Congress? And, does this not open the door for criminal activity or the unauthorized use of taxpayer money?

Why was the Nazi SS General Gehlen allowed to continue independently operating his spy organization in Eastern Europe until the 1960s, when it was incorporated into the West German BND intelligence agency?

Why did the CIA, at one time after the World War II, have an estimated 50% of Nazi SS employees?

What was the full nature of the secret treaty agreed to between President Eisenhower and representatives of the Nazi International?

Is it possible that U.S. citizens are not allowed to know the answer to these questions because the Nazi International now controls the U.S.?

The Second Secret Space Program

While the Germans were developing their secret space program, interesting things were also happening in the U.S. in the early part of the Twentieth Century.

In January 1901, the Martians sent a 70 minute signal to Earth that was picked up by the Lowell Observatory.

While Nikola Tesla was conducting experiments with his Magnifying Transmitter at Colorado Springs in 1899, he detected coherent signals which he determined had originated on Mars. Tesla was widely criticized for his astounding claims, yet no one could seriously dispute him, for he was a solo pioneer without peer.

About 40 years later, Arthur Matthews claimed that Tesla had secretly developed the "Teslascope" for the purpose of communicating with Mars. Dr. Andrija Puharich met with Matthews and discussed the matter with him in an interview which was published in the *Pyramid Guide*:

Arthur Matthews came from England. Matthews' father was a laboratory assistant to the noted physicist Lord Kelvin back in the 1890s. Tesla came over to England to meet Kelvin to convince him that alternating current was more efficient than direct current.

Kelvin at that time opposed the AC movement. But later, after hearing Tesla out became convinced of the superiority of using AC. In 1902, the Matthews family left England and immigrated to Canada. When Matthew's was 16 his father arranged for him to apprentice under Tesla. He eventually worked for him and continued this alliance until Tesla's death in 1943.

According to Matthews:

> "It is not generally known, but Tesla actually had two huge magnifying transmitters built in Canada, and Matthews operated one of them. Most people know about the Colorado Springs transmitters and the unfinished one on Long Island. I saw the two Canadian transmitters. All the evidence is there.
>
> The Teslascope is the thing Tesla invented to communicate with beings on other planets. There's a diagram of the Teslascope in Matthew's book, *The Wall of Light*. In principle, it takes in cosmic ray signals. Eventually the signals are stepped down to audio. Speak into one end, and the signal goes out the other end as a cosmic ray emitter."

In the 1920s and 1930s Martians organized tours of Mars for Earth elites that were conducted through the Vatican with pick-up destinations at known locations including the southwest USA and Brazil.

By 1970, Mars astronauts were conducting regular liaison visits respective countries and used the public consciousness of the cold war to have a competitive "space race". But this race would only use already obsolete rocket technology.

At the same time both countries were cooperating in a joint secret space program, which would utilize the captured German antigravity technology.

It is estimated that this joint secret program was about 20 years in research and development behind what the Germans had already accomplished.

Some of the financing for this secret space program came from creative bookkeeping of the funds for the official space programs in each country.

A large motivation for this secret joint space program between the Soviets and the U.S. was that both nations feared an alien extraterrestrial attack on earth. The U.S. passed very strict laws and protocols, protecting the secret of extraterrestrials, at first to keep the public from panicking, and later to keep secret technology received from extraterrestrials from public view.

The recovering of crashed flying saucers led to a program of back engineering this exotic technology. It was so advanced that the Project Paperclip Nazis had to be used to do the back engineering. President Truman placed this program under the management of MJ-12.

Originally, the back engineering was done at Wright Patterson Air Field, but was later moved to area S-4 at Area 51 of the Nellis Test range in Nevada for greater secrecy.

The Nazi International or ODESSA was very much aware of

these developments and was manipulating both sides of the "cold war" for their own advantage, while watching the progress of the joint secret space program via their own German spies within the program.

Eisenhower, following the recommendations of Nelson Rockefeller, placed MJ-12 under the control of the CIA. This would keep future Presidents "out of the loop" and insure secrecy. The Rockefellers had been financiers of both the Bolsheviks and the Nazies, as so well documented by historian Anthony Sutton. And, the CIA was created by Knights of Malta agents of the Vatican.

So secret agents of the Vatican through the CIA, actually had control over MJ-12. The Vatican was the core of the Nazi International. And the Rockefellers assisted in their financing. So here we see why it was so easy for the Nazi International to infiltrate the highest levels of the military industrial complex involved in "black projects" of the United States.

Later, however Eisenhower began to regret this decision. Now, even he was in the dark about what was going on at Area 51. He even went so far as to threaten a military invasion of the secret base to find out what was going on. He really didn't trust the MJ-12 group or the CIA.

He also created a secret division of the Marine Corps called Special Section to have their own secret space program which would be independent of MJ-12. Eisenhower felt that the loyalty and the integrity of the Marine Corps officers would be more trustworthy than the MJ-12 and the CIA. And, that they would be more likely to uphold the Constitution during these unsettling times.

Article 21 of secret Marine Corps regulations authorizes the US Marine chain of command to deploy a U.S. Marine as a civilian to speak out publicly against the actions of the U.S. secret government when either of two conditions is reached:

- Less than 50% of the constitutional guarantees of the US government are in place.

- Less than 75% of the operating functions of the US government are no longer being carried out.

Eisenhower placed Article 21 into the Special Section code of conduct because he feared the Nazi infiltration of the Military Industrial Complex and what that bode for the future of the U.S. On parting from the Presidency, Eisenhower's farewell speech included a warning about the Military Industrial Complex which he realized was being taken over by the Nazi International.

Further on, we will see where this Article 21 has been activated by some high officers of the Marine Corps, Special Section because of the sorry state of the present U.S. government.

And, because secrecy has become a greater threat to the Constitutional government of the U.S. than any conceivable enemy is.

By the 1960s the U.S and Russia had their own operational flying saucers. Because of the secret treaty between Eisenhower and the Nazi International, they were allowed to fly a joint U.S. - Russian saucer to the German moon base and were taken on a tour.

Later the U.S. gained control over this Moon base. Corey Good explains how this came about:

"It wasn't that the Nazi base that became the LOC was abandoned. During the 1950's and after they had successfully infiltrated and subverted the Military Industrial Complex and major Corporate heads they had effectively won control of the direction of not only the Breakaway Civilization Programs but also the mainstream government and financial system.

It was a very effective and silent coup that gutted what was once the American Republic and turned it too into a Corporate Entity with each of us being "Assets" with our very own serial numbers.

This plan was in action far before World War One by various secret societies who controlled the financial system and as many know financed both sides of the wars.

When both Truman and Eisenhower signed treaties with the NAZI Break Away Civilization/Societies it was then that the already well placed Operation Paperclip Operatives (in Military, Corporate Industry, Intelligence and established Secret and Public Space Programs) easily slid into more powerful and influential positions over the massive industrial complex of the USA that they coveted to expand their operations in space and form what would later become the ICC (as well as were involved in setting up all of the other Space Programs).

So they did not lose the Lunar Base, they infiltrated the groups that put a massive effort into expanding it into the massive complex that it is now."

In May of 1962, a joint U.S. - Russian flying saucer lan'ed on Mars! Bulgarian researcher from the Russian Academ

Science, Vladimir Terziski, supplied me, in 1989, with a VHS video showing this landing. The video was an excerpt from the video *Alternative 3*.

The craft was banking and turning through Martian valleys, searching for a good landing zone.

The video was taken through a curved window, as would exist on the saucer shaped craft which was definitely not rocket powered. In the background could be heard excited voices in English and Russian.

After the craft landed, atmospheric temperature and pressure was measured. It was equivalent to that of Earth on a high mountain.

At the end, the video showed some form of burrowing life that was moving. The same video that I saw is now available on the internet here: http://www.bibliotecapleyades.net/luna/esp_luna_46.htm

The Germans had gone to Mars earlier than 1962 and also created bases on Mars. But, many of the the early bases were attacked by indigenous Martians and were wiped out. It took a while for them to establish permanent bases, this was after they had taken over the U.S. military Industrial Complex. Actual colonization of mars started in 1964 according to Terziski. Many of the more educated people of Britain and the U.S.were secretly relocated to these colonies.

They were informed that the Earth would soon suffer a tremendous calamity and the Mars colonies were part of a project to preserve mankind. This resulted in the "brain drain" of the 1960s through the 1980s mentioned in *Alternative 3* - a revelation

shape with the diameter growing size as you go down. Goode only had access to the 4 top levels, but saw plan of the base.

The base is divided between regular military personnel and VIP areas. There are small rooms for military personnel and the walls are colored blue or gray. The VIP areas are larger and have beautiful wood veneer walls.

There is definite compartmentalization of access to different portions of the base, which is run in a strict military fashion. There are office spaces and also conference rooms where delegates from different star systems and Earth leaders meet.

These conference rooms are also used to give speeches to military personnel when beginning a mission or ending a mission. Randy Cramer remembers that his end of service speech was given by Donald Rumsfeld at the LOC conference room.

The transport craft are triangular shaped, can carry up to 300 passengers and can traverse space from the Moon base to Mars in about half an hour and to Earth in much less time. So, here we see just how obsolete rocket propulsion really is.

In *Secret Science and the Secret Space Program*, I explained how these craft were created in "black projects" by military defense contractors like Lockheed Martin, Northrop Grumman, Boeing and others.

Corey Goode also explains that the Moon is used as a diplomatic meeting ground for a number of extraterrestrial civilizations and a number of other extraterrestrial bases are located there, primarily on the far side of the Moon, in zones divided up like Antarctica is.

The Mars colonization was secretly carried out by multinational corporations in conjunction with the military and their Solar Warden program. Randy Cramer stated that It was called the Mars Colony Corporation where he served for 17 years of his 20 year tour of duty. Corey Goode claimed it was the Interplanetary Corporate Conglomerate (ICC), indicating that there were other colonies besides on Mars. Apparently, the Mars Colony Corporation is a subsidiary of the ICC.

The people in many of these colonies were totally cut off from any news from Earth and were told that a major catastrophe had ruined civilization back on Earth and there was no hope of ever returning to their home planet. Although fed,clothed and housed,these colony workers were kept in almost slave labor like conditions and were controlled via propaganda.

These conditions however were not as bad as the prison laborers of the Third Reich in which many of the workers died from horrible living and working conditions. Corey Goode traveled to these colonies to make emergency repairs and was given strict orders not to make eye contact with anyone nor speak to anyone at the colony about anything not related to the repair job.

The Nazi International that infiltrated the U.S. aerospace industries - like Northrop Grumman, Lockheed Martin, Boeing Hughes Aerospace, and General Electric, whose black projects created the technology for the second Secret Space Program - also brought their racial eugenics ideas with them.

They were selectively targeting certain genetic parameters in persons picked for the Mars and other colonies (including Alternative 1, deep underground colonies). These people were definitely high IQ Aryan type people which caused the

Alternative 3 "Brain Drain" in the 1960s - 1980s.

Actually limited recruiting for the Mars colony is still ongoing. One person, with the right bloodlines and IQ, targeted for the mars Colony was the great granddaughter of President Eisenhower, Laura Magdalene Eisenhower.

Laura Magdalene Eisenhower, who has resided and traveled independently in over 20 cities in the U.S. and abroad, has developed a wide knowledge base in frontier health, natural systems, alchemy, metaphysics and ancient history, and also has degrees and certifications in science, wilderness expedition leadership, natural healing and building. She is a mythic cosmologist, global strategist, clairvoyant healer, Earth advocate, and artist.

The method of recruitment was to set up a love affair between Laura and the mind controlled lover they selected to capture her heart and mind. This is what she had to say:

> "He knew a lot about me from numerous sources that overlapped with one another – from Freemasons, Knights Templar, to this hidden branch of the government that was behind creating this Mars mission.
>
> They understood me based on remote viewing and time travel devices and they also seemed to recruit people from the Freemasons and Knights Templar who were well aware of the Magdalene path.
>
> They seemed to all be connected, but someone was taking advantage of it all and trying to stay one step ahead of how nature may have unfolded as far as us coming together into union.

I did not know he was an agent until months into the relationship and I later found out that it was them who sent him to find me and it was all a set-up."

It is interesting to note the Freemason and Knights Templar connection to the Mars colony.

Many researchers including myself have traced these secret societies back to the Jesuit order that now controls the Vatican and is in back of the "New World Order". The Magdalene path, named after Mary Magdalene, is to awaken the feminine soul to it's higher possibilities.

The idea was that Mars would have the ideal genetic population. Then, the surface population of the Earth would be wiped out via war, biological plagues, chemtrail poisoning and genetically modified organisms.

After the "cleansing" of the surface populations from the Earth, the ideal genetic people from these various colonies would re-populate the Earth and the population would be carefully controlled to be kept under 500 million people as indicated in the Georgia Guidestones.

Apparently, after certain high level leaders in the military discovered the whole plan and felt that it constituted a tremendous crime against humanity, as well as a very real threat to themselves and family, they decided it was time for the secrecy to end.

Capt. Randy Cramer states straightforwardly that he was authorized under Article 21 by his chain of command to speak out. He says,

"When I agreed to speak publicly, my security clearance was raised to a Blue/Gold-13, which has granted me full access to USMC ss intelligence files, and weekly briefings by Brigadier General Julian Smythe, personally."

According to Capt. Cramer,

"USMC Special Section (ss). is a covert Unacknowledged Special Access Program (or USAP) signed into law as a legal and covert branch of the US military in 1953 by President Dwight D. Eisenhower. USMC ss was mandated by President Eisenhower through the USMC ss special code to meet the exopolitical questions of EBE's, (Extra Terrestrial Biological entities) and ETV's (Extra Terrestrial Vehicles); to assist in assessing diplomatic opportunities and military threats and advise the MJ-12 committee and special study groups (SSG's) with intelligence from a fully staffed and operational military intelligence machine, and respond to their requests for specially trained military assistance in all matters extra terrestrial."

Capt. Randy Cramer has a two-fold message to the American and by extension the world public. The American and Earth's surface dwellers have been left with an archaic economy based on outdated technology by a Breakaway civilization whose technology is 10,000 years or more advanced.

Only by securing release of secret green, advanced energy and other technologies can Earth's surface dwellers restart their economy and thrive.

The Breakaway civilization is preventing this release of technology (like anti-gravitics, time travel teleportation, medical and other technologies). Their plan, called Alternative 3, is based on a secret gene pool Mars colony established to repopulated Earth after the Earth surface dwellers are eliminated.

The means of elimination include intentional biowarfare, wars, tectonic and environmental wars (HAARP and chemtrails), GMOs and other toxic means. It is disinformation to believe that Alternative 3 was established in the event of a natural ca-tastrophe on Earth. Alternative 3 was the planned destruction of surface humanity.

Capt. Randy Cramer has now been ordered by the US Marine Corps chain of command to warn the American and Earth pop-ulation this is occurring. The US government is complicit in this plan and can no longer carry out its function of protecting its citizens.

Capt. Randy Cramer 's solutions include a call for re-energize citizen participation in oversight of official secrecy in the mili-tary-industrial complex.

Capt. Cramer encouraged citizens to become actively involved in elections, and oversight of Congressional representatives and government offices.

Asked if he planned to run for elective office as part of his man-date under Art. 21, Capt. Cramer replied that he was consider-ing political campaigning for public office among other options.

It is not clear if the US Marine Corps has other robust options available to it to preserve the American nation and people un-der Article 21 of its secret regulations, should the challenges to the continuity of the constitutional integrity of the American people's nation become even more challenging.

Besides Randy Cramer and Corey Goode, there have been a number of whistleblowers that have come forward and revealed portions of these Secret Space Programs.

Lawyer, Andrew D. Basiago is admitted to the Washington State Bar Association and the United States District Court for the Western District of Washington.

He served in DARPA's Project Pegasus from 1968 to 1972 and the CIA's Mars Project from 1980 to 1984. Project Pegasus involved teleportation and time travel on this planet. The Mars project involved Teleportation to Mars.

When Randy Cramer first came forward to reveal his role in the Mars Defense Force, Basiago compared his story, to his own experiences on Mars and pretty much said that Cramer's story was probably true while many others were saying that Randy Cramer was crazy.

Dr. Stephen Greer organized the Disclosure Project, organizing whistleblowers that worked on secret military programs with secret technology.

If anyone in these covert programs decided to become a "whistleblower" the usual procedure is to make all records of that person, birth certificates, educational records, employment history Etc. "disappear." Usually this means that person will suffer an end to their career. And sometimes whistleblowers end up "suicided" or just "disappear" themselves. So, becoming a whistle blower is not an easy decision for these people to make. Otherwise, there would probably be a lot more of them.

In one example of this secrecy, Vice Admiral Tom Wilson was J-2 head of intelligence for the Joint Chiefs of Staff at the Pentagon.

Dr. Steven Greer of the Disclosure Project turned over some documents about classified extraterrestrial related projects which included a list of code names and project names over to Wilson. When Wilson checked to determine if the projects really existed, he was denied access.

Greer said that when Wilson identified the group, he told the contact person, "I want to know about this project." And, he was told, "Sir, you don't have a need to know. We can't tell you." The contact persons were not DOD. They were attorneys for the defense contractors! (8)

So, here we see an example of what President Eisenhower was warning the American people about in his farewell speech. The Military Industrial Complex, or parts of it, that neither the President nor the Pentagon have knowledge of or control over!

The Solar Warden program started in the 1980s and was the secret part of the Strategic Defense Initiative called for by President Ronald Reagan.

According to their web site, the United States Space Command (USSPACECOM) is a Unified Combatant Command of the United States Department of Defense, created in 1985 to help institutionalize the use of outer space by the United States Armed Forces.

The U.S. Space Command is headquartered at Peterson Air Force base at Colorado Springs, Colorado at the following address and Phone number:

Air Force Space Command
Public Affairs Office
150 Vandenberg St., Suite 1105

Peterson AFB, CO 80914-4500

(719) 554-3731 or DSN 692-3731

The U.S. Space Command had its computer database hacked in 2002, by British Citizen, Gary McKinnon, who discovered on the database a list of "Non-Terrestrial Officers" and " fleet to fleet transfers." The names of the ships listed on the "fleet to fleet transfers" did not correspond to any U.S. Navy ships.

McKinnon also stated that the records revealed that the off-planet shuttles could hold 300 people. That statement is reinforced by a statement from President Ronald Regan's diary. This diary entry for Tuesday June 11, 1985 reads:

"I lunch with 5 top space scientists. It was fascinating. Space is truly the last frontier and some of the developments there in astronomy etc. are like science fiction, except they are real. I learned that our shuttle capacity is such that we could orbit 300 people."

The space shuttle that the public knew about could only hold 8 people. So obviously there was one that was secret that Reagan knew about and McKinnon discovered that could hold 300 people. And you can bet that it used antigravity technology.

Reagan also made an interesting speech at the 1985 U.N. General Assembly:

"In our obsession with antagonisms of the moment, we often forget how much unites all members of humanity. Perhaps we need some outside, universal threat to make us recognize this common bond. I occasionally think how quickly our differences worldwide would vanish if we were facing an alien threat from outside this world. And yet I ask- is not an alien force already among us?"

This reference to aliens was not on the speech his staffers had approved. So, Reagan ad libbed this speech. And, what did he mean about an alien force already among us? Did Knight of Malta, Ronald Reagan have access to information usually classified many levels above what the President of the U.S. is allowed to know?

Interestingly Gorbachev responded to this speech at the 1985 Geneva Summit:

"At our meeting in Geneva, the U.S. president said that if the earth faced an invasion by extraterrestrials, the United States and the Soviet Union would join forces to repel such an invasion. I shall not dispute the hypothesis, although I think it's early yet to worry about such an intrusion."

But, from many other sources it appears that the U.S. and Soviets were already cooperating in Space. Perhaps the Cold War, like NASA was also a smoke screen.

Does McKinnon's discovery mean the Space Command has fleets of spaceships in space that Non-Terrestrial Officers get transferred between in fleet to fleet transfers? And, more importantly, when do these Non-Terrestrial Officers get " shore leave" back on Earth?

McKinnon also hacked into NASA and Pentagon computers where he found film, photographs and other evidence of spacecraft secretly held by these agencies. One picture in space showed a high definition photograph of a large cigar shaped object over the northern hemisphere of planet Earth.

This was most likely a Solar Warden mothership that was developed from the German cigar shaped, *Andromeda* space going, "aircraft carrier" that carried their flying saucers inside.

A good description of these motherships, provided by Randy Cramer, who served 3 years on one, in an interview with Michael Salla, is given in my *Secret Science and the Secret Space Program* in Chapter 9, Beyond Antigravity.

In answer to the question, "when do these Non-Terrestrial Officers get " shore leave" back on Earth?" , according to Michael relfe, Randy Cramer and Corey Goode – they don't. At least, not until their 20 year tour of duty is completed.

When this happens, they are taken back to LOC on the moon and have their memories of their 20 year service wiped electronically from their minds. At the same time, they are they are age regressed by 20 years using advanced alien technology and then sent 20 years back into the past with their time control technology.

To them, it is almost like their 20 year tour of duty never even happened! They can continue living a normal life back on earth. At least they can in theory. The human mind works in strange ways. The subconscious mind retains all intense feelings and trauma - even when the conscious mind has forgotten. This can lead to Post Traumatic Stress Disorder (PTSD) in combat veterans.

The electronic memory wipe of the conscious mind didn't affect the subconscious mind. Both Michael Relfe and Randy Cramer had seen violent combat on the Martian surface. Also Randy Cramer had fallen in love with a female trooper of the Mars Defense Force and married her. Often they would fight side by side in defending the Mars Colonies. In one battle, his wife was killed and it saddened Randy Cramer considerably.

Randy Cramer recalls that before getting his memory wipe at the LOC, how badly he wanted to remember his dead wife. Both men suffered from PTSD after returning to Earth.

Michael Relfe got married to a neural linguistic and biofeedback therapist. She started giving him therapy for his PTSD and gradually he was able to recover his lost memories. Michael Relfe's therapy and recollections of his Mars duty are recorded in *The Mars Records* written by his wife Stephanie Relfe. Here is part of his story:

> "I enlisted in the Navy in 1976 for the Nuclear Power Program. That program requires a six year enlistment. My time in the Navy was 6 years relative and 26 years absolute.
>
> That means 6 years in my "normal" life timeline. Then some time after arriving at Great Lakes, Illinois for Electronics Tech school I was recruited to Mars.
>
> I stayed there 20 years, at which time I was age regressed and returned to Great Lakes about a week after I had left.
>
> Remember that during the duration of my 6 year enlistment plus the next 14 years, there were two instances of "me", one on Earth and one on Mars. There was no "parallel universe" situation.
>
> In the book [The Mars Records], the visual representation of the timeline helps clarify the matter.

I was living my normal "timeline" when I was recruited for the program. I was taken to Mars via jump gate where I served a 20 year tour of duty. During that period I was not allowed contact with any event, anything or anyone on earth.

At the end of the tour of duty I was physically age regressed 20 years and returned to my point of origin. (about a week after I had left). I continued to live my normal "timeline"

From the point of an outside observer I lived my normal time line (minus the week I was gone). At the same period I was also living on Mars, totally isolated from earth. The key is ISOLATED. During the time on Mars I was never allowed to interfere with any event on earth. In addition, I eventually had to be returned to earth to live out my normal timeline and accomplish my destiny."

The jump gate Michael Relfe referred to was developed from the Philadelphia Experiment and the Montauk project that Al Bielek describes. The technology had been well developed by the time Michael Relf was enlisted and was used as an everyday practice .

Michael also clarified that while he had to spend 20 years on Mars, VIPs from Earth would come in with these jumpgates and stay a week or two and then be returned the same way without memory wipes, age regression or time travel.

Six independent whistleblower witnesses, including Michael Relfe, have confirmed the Jumpgate technology and the existence of one or more U.S. secret bases on Mars, as forward strategic military bases for occupation or defense of the solar system.

These whistleblower witnesses include, besides Michael Relfe, former U.S. Army Command Sgt. Major Robert Dean, former participant in DARPA's Project Pegasus Andrew D. Basiago and former U.S. Department of Defense scientist Arthur Neumann, Member of the Marine Corps Special Services, Randy Cramer and Solar Warden research vessel crew member Corey Goode.

Andrew D. Basiago is a former participant in DARPA Project Pegasus (1968-72) that developed Tesla-based quantum teleportation and time travel in the time space hologram, initiating the U.S. program of time-space Chrononauts. According to Mr Basiago, the U.S. government already had a fully operational teleportation capability in 1967-68, and by 1969-70, was actively training a cadre of gifted and talented American schoolchildren, including himself, to become America's first generation of "chrononauts" or time-space explorers.

Mr Basiago has revealed that between 1969 and 1972, as a child participant in Project Pegasus, he both viewed past and future events through a device known as a "chronovisor" and teleported back and forth across the country in vortal tunnels opened in time-space via Tesla-based teleporters located at the Curtiss-Wright Aeronautical Company facility in Wood Ridge, NJ and the Sandia National Laboratory in Sandia, NM.

The Wood Ridge NJ, Curtiss-Wright Aeronautical Company facility was also the location Andrew Basiago witnessed Martian Astronauts with his father.

A chronovisor is a device that uses a screen or holographic template to locate and display scenes from the past or future in the time-space hologram. The chronovisor was back engineered extraterrestrial technology by two Vatican scientists in conjunction with Enrico Fermi and later refined by DARPA scientists.

DARPA had, he explains, five reasons for involving American schoolchildren in such new, dangerous, and experimental activities:

1. First, the Department of Defense wanted to test the mental and physical effects of teleportation on children.

2. Second, Project Pegasus needed to use children because the holograms created by the chronovisors would collapse when adults stood within them.

3. Third, the children were tabula rasa (of the mind that has not yet gained impressions of experience) and would tend to see things during the time probes that adults would tend to miss.

4. Fourth, the children were trainees who upon growing up would serve in a covert time-space program under DARPA that would operate in tandem with the overt space program under NASA.

5. Lastly, the program sponsors found that after moving between timelines, adult time travelers were often becoming insane, and it was hoped that by working with gifted and talented children from childhood, the U.S. government might create an adult cadre of "chrononauts" capable of dealing with the psychological effects of time travel.

This training, he said, culminated in 1981, when, as a 19-year-old, he teleported to Mars, first by himself after being prepared for the trip by CIA officer Courtney M. Hunt.

And then a second time in the company of Hunt. Both trips, Mr

Basiago said, were made via a "jump room" located at a CIA facility in El Segundo, CA.

The apparent purpose of the trips to Mars was to familiarize him with Mars because the CIA knew of his destiny pertaining to publicly establishing the fact that Mars is an inhabited planet and deemed it important that he visit Mars and experience its conditions first-hand.

Mr Basiago's involvement in advanced U.S. time-space research as a child, as well as Courtney M. Hunt's identity as a career CIA officer, have been confirmed by Dr. Jean Maria Arrigo, an ethicist who works closely with U.S. military and intelligence agencies, and by U.S. Army Captain Ernest Garcia, whose storied career in U.S. intelligence included both serving as a guard on the Dead Sea Scroll expeditions of Israeli archeologist Yigal Yadin and as the Army security attaché to Project Pegasus.

After returning to Earth. Randy Cramer sought Yoga and meditation as a way to overcome his PTSD symptoms. Gradually, bits and pieces of his memory, especially of his wife, started returning. At first he thought that he was going crazy, as nothing made logical sense and because he had no conscious knowledge of any Mars colony.

However, total recall of his memories finally returned when his two timelines came together after 20 years back on Earth. It must cause certain psychological stress to be in two places at the same time. After 20 years, Randy Cramer's life started to become more normal.

Randy Cramer described being taken to the Lunar operations Command (LOC). On November 17, 1987 at 2:30 A.M., a wormhole materialized in his room, he was taken to an underground base via this wormhole and placed on a TR-3B on which he was flown to the LOC.

He is quite an interesting person and has an informative website here: http://www.earthcitizenconsulting.org/

Corey Goode, likewise did the standard 20 year tour of duty, which he signed up for at the LOC. He was stationed on a research vessel and his experiences were much less traumatic.

During this time he had access to "Smart Glass" pads, much like an iPad. These "Smart Glass" pads were an encyclopedic source of all kinds of information within the Secret Space Program.

Because of this access to these "Smart Glass" pads, Corey Goode was able to learn a lot about these programs that most personnel had no access to.

He likewise, had his memory wiped, was age regressed 20 years and time traveled back to the time of his enlistment. His story is quite involved and lengthy. So I won't cover too much of it here, although he is an excellent source of information which is used in this book elsewhere.

Much of his story is revealed on David Wilcox's video interviews with Corey Goode at GaiamTV in the Cosmic Disclosure series. It is quite interesting and enlightening. Here is a U-tube frce sample: https://www.youtube.com/watch?v=ukpPtzvracg

I will say that one thing that Corey Goode emphasizes in his talks is a subject he calls "complete disclosure", which he claims will soon come to this planet.

The people of the world have been kept completely in the dark about these Secret Space Programs and the technology used to make them happen. Much of this technology would, if put into use on this planet, would dramatically change everything.

For example, the technology of free energy, where energy is extracted directly from space itself, would make it possible for people to have independent home power units with no power bills and no ugly power and transmission lines needed. There would be no power outages when storms bring power lines down. Cars would all become non polluting electric cars with infinite range and no recharging time. Airplanes could lower air fairs because they would have no fuel costs. The economy would immediately improve, as transportation costs would dramatically lesson and people would have more money to spend that they would have previously wasted on and electric energy and fuel costs.

The antigravity technology would allow people to have their own flying saucers with infinite range that would allow them to travel anywhere on the planet or with more elaborate craft, anywhere in the solar system. People with a tight schedule could go to the local teleport station and travel immediately to their desired destination.

The reason this science and technology has been withheld from the common people of Earth is to keep them in economic bondage to a cabal of elite who derive their income from selling fuel, creating unnecessary wars, and keeping everyone drugged up and uninformed about what is really going on.

This elite think that by keeping everyone at each other's throat, they will be keeping them away from their throats.

The present health care system in the U.S. is a total racket. I have seen hospital costs increase over 1,000% since the 1960s for the same services. Doctors over subscribe medications unnecessarily so that the pharmaceutical companies can make fortunes. Health insurance is also a big money maker.

I spent most of my life without health insurance and would just pay cash on the rare occasion of visiting a hospital. But nowadays the astronomical health care costs and the Obama Care law, require me to have it. Luckily, I qualify for Medicare.

The medical technology used in the Secret Space Programs would revolutionize health care and allow people to live much longer and youthful lives with little expense.

Corey Goode states that full disclosure will awaken the people to the true state of affairs and the true history of the last several centuries. Many government and military officials will be on trials for their crimes against humanity.

Bankers will likewise be prosecuted for their criminal monetary manipulations. Money will become much less important and governments will create their own money, in place of the present debt based system designed to keep most of us in economic bondage.

A golden age of peace and prosperity for humanity will ensue as the spiritual development of mankind accelerates and mankind demonstrates their ability to handle their greater abilities with harmony and responsibly. People will not be barraged with scenes of violence from the news and entertainment industry, which presently is being used to program our minds towards fear, greed, lustful and violent tendencies rather than to elevate consciousness towards trust, peace and love for one another as true religion has taught and which will make us truly happy.

When one judges someone else, it usually is what that person themselves would do. Judgment of people with different racial, religious, economic, or ethnic backgrounds as being somehow less than ourselves, limits our own ability to truly enjoy these differences for what they truly are, and learn a different point of view. After all, we all are human beings. People that are well traveled know what I am saying here. Judgemental people are usually a bore to be around.

Imagine what it is going to be like when we become interstellar travelers and get to experience all those other civilizations out there.

Cory Goode says that our civilization will eventually become like a "Star Trek" civilization after full disclosure.

Global Galactic League of Nations

Corey Goode reveals the extent of the global conspiracy to conceal the secret space programs through a multinational group known as the Global Galactic League of Nations.

The United States was retrieving crashed flying saucers in many different countries and treaties were established to allow this to happen. It didn't take these countries long to find out about the secret U.S. - Soviet space program and it's purpose to defend the planet from a feared extraterrestrial attack. These smaller countries demanded to be involved in something this important. The U.S. and Soviets still wanted to maintain secrecy. So their problem was what to do about these smaller countries.

According to Corey Goode in the 1987, they organized a secret Global Galactic League of Nations as a type of "carrot" to offer to these smaller countries in the form of membership in this organization. A pledge of allegiance and secrecy was required to become a member.

Many of these smaller countries wanted to get on board this important program to protect the planet. It made them feel more important.

As an added inducement, space bases and technology was also offered these member countries. However for whatever reason, these gifted bases were located outside our solar system. Perhaps it was to keep secret the colonies using slave labor within the Solar System.

Corey Goode himself was at one of these extra solar system bases, which he described as being much more relaxed and easy going than the militaristic run Mars Colony he had visited.

These distant bases were traveled to using portal systems the Secret Space Programs had learned to use and could be traveled to very quickly. These bases were filled with many research scientists and had all the latest technical "toys" according to Goode.

Also according to Goode, these people had little knowledge of the space programs going on within the solar system!

The particular base that Corey Goode described, was on a moon surrounding a gas giant and had had a breathable atmosphere. It was located within a huge cave with a unsealed entrance to the rest of the moon. Some of the personnel were frolicking in a beautiful lagoon within the cave in their time off.

The history of the more international space program started with Solar Warden, which started in 1962, with secret meetings between the Star Nations and 4 major Earth Governments, The U.S., Russia, China, and the European Council (forerunner to the European Union).

At that time MJ-12 was internationalized under the United Nations Security Council and the original MJ-12 became the U.S. Special Studies Group. That way, the U.S. wouldn't have exclusive control over ET related matters.

The current (2014) members of this international MJ-12 are:

MJ1 - Simonetta Di Pippo, the Italian director of UNOOSA.

MJ2 – Brigadier General Gartzene Killennarry, senior manager at British MI-6.

MJ3 - Anke Schaferkort, German director of BASF.

MJ4 – Hon. Paul Hellyer, former Canadian Minister of Defense.

MJ5 – Samantha Power, US Ambassador to the UN.

MJ6 – Thai Male, Professor of Software Engineering at Bangkok University's School of Science and Technology.

MJ7 – Nelson Violante Carvalho, Ph.D. tenured lecturer at the Federal University of Rio de Janeiro.

MJ8 – Olga Golodets, Deputy Chairman for Social Affairs , Federal Cabinet of the Russian Federation.

MJ9 – John Chandler, Commander of the Space and Naval Warfare Systems Command.

MJ10 – Joyce Victoria Bigio, Italian Chief Liaison to the Vatican.

MJ11 – Carmen Omonte Durand, Peruvian Congresswoman

MJ12 – Dr. Ross McKenzie, professor of Quantum Physics at University of Queensland.

MJ-12 has been directed in recent times by an Executive

Committee composed of CEO Simonetta Di Pippo(MJ1), Samantha Power (MJ5), Joyce Victoria Bigio (MJ10), and Captain John Chandler, USN (MJ9).

Named Director of UN Office of Outer Space Affairs (UNOOSA) in March, 2014, Simonetta Di Pippo is overseeing the gradual integration of MJ-12's Executive Committee into UNOOSA as an executive advisory committee. This is being done because soon there will be no need for a UFO secrecy-management organization.

After integration, members of MJ-12 will be converted into advisors for UNOOSA as needed. Ms. Di Pippo is working with Dr. John Ashe, UN President, to increase the democracy of UN government.

This is done by strengthening the role of The General Assembly and the UN President, and revising the obsolete, oligarchical arrangement of the Security Council's making all the major decisions.

In 1964, a secret meeting took place in an underground base under El Capitan Peak in northwest Texas. The meeting was the culmination of a series of meetings first initiated by President Eisenhower in 1960.

The U.S. was represented by David Rockefeller, the USSR by Nikita Khrushchev, China by a trusted associate of Mao Tse-Tung and the European Council by Anthony Bradley. The Star Nations representative was Asheoma Maeta.

The Star Nations concerns were the irresponsible testing of nuclear weapons and the great harm to the Earth these tests were causing. The nuclear explosions also were causing tears in the space time continuum and were causing problems for the beings on other dimensions. Another concern was the shooting down of their craft by the military of different Earth governments using their new high energy pulsed radar systems.

The 4 Earth Nations representatives wanted the Star Nations to give them more ET technology including antigravity and genetic technology in return for a promise of a cease fire. They also wanted the ETs to restrict how close their mother ships could approach the Earth to stay out of view of the Earth people. The scout craft were limited to how close they could fly to the Earth unless on a mission pre-announced to the government. They also wanted the ETs to not interfere in any of the Earth's religions or commercial arraignments. Finally a treaty was agreed to and signed in 1964. After this, private contactee meetings between private people and ETs became considerably reduced.

After decades of violations of the treaty between the U.N. and the Star Nations, the worst being the shooting down of Star Nation spacecraft, the 1964 treaty was deemed null and void by the Star Nations.

Finally, it was decided that a certain faction was responsible for these treaty violations. This faction was identified as controlled by the Jesuit order of the Vatican, whose secret societies had infiltrated the banking, energy, military and political segments of many societies. They feared the ETs because they could threaten their economic, military and political control of the world's people.

This faction became known as "The Cabal". Known leaders of this Cabal include the Rockefellers, the Rothschilds, Henry Kissinger, Prince Bernhard of the Netherlands, George Herbert Walker Bush, Pope Pius XII and many more of similar ilk who have caused so many unnecessary bloody and cruel wars on our planet.

Finally, negotiations were continued with the understanding that the government leaders did not truly represent the will of the people of Earth. The UN Committee on the Peaceful Uses of outer Space was transformed into the Outer Space Affairs Division in the UN Security Council in 1968. It was transformed again into The United Nations Office for Outer Space Affairs (UNOOSA) in 1993, whose stated purpose is "promoting international cooperation in the peaceful use of outer space." At that time, the office was relocated to the UN office in Vienna, Austria.

Solar Warden is part of a secret extraterrestrial treaty agreement with the Star Nations – the organization of advanced intelligent civilizations in space – and the United nations.

Because of its advanced technological position, the U.S. has been designated by Star Nations to a lead position in providing space security for Earth. The U.S has been adding to the Solar Warden fleet since the late 1980s using black projects within the aerospace corporations.

But, the U.S. is not the only nation involved. Other nations with advanced technologies are also involved including Canada, United Kingdom, Italy, Austria, Russia, and Australia. Also, there is a U.N. authority involved, the United Nations Office for Outer Space Affairs (UNOOSA). While the majority of the people staffing the mother ships and scout ships of the Solar Warden Space Fleet are Americans (U.S. Naval Space Cadre), there are also some crew members from UK, Italy, Canada, Russia, Austria, and Australia.

The mission of Solar Warden is twofold:

One part of the Space Fleet's mission is to prevent rogue countries or terrorist groups from using space from which to conduct warfare against other countries or within-country targets. Star Nations has made it quite clear that space is to be used for peaceful purposes only.

The second part of the Space Fleet's mission is to prevent the global-elite control group, the Cabal, from using its orbital weapons systems, including directed-energy beam weapons, to intimidate or attack anyone or any group it wished to bend to its will.

Because the Space Fleet has the job of being Space Policeman within our solar system, its program has been named Solar Warden.

Star Nations has not given the U.S. Government exclusive authority to police the Earth. The U.S. has no authority from Star Nations to engage in any international policing activities. Star Nations has the policy position that the citizens of Earth have the responsibility to work out the operation and regulation of their societies as best they can.

The mandate and jurisdiction of the Solar Warden Space Fleet is space. It does not have jurisdiction and does not meddle in Human affairs on the ground, nor Human activity occurring within Earth's atmosphere. Those are the jurisdictions of the respective governments in each country and the air space above their territories.

The Solar Warden Space Fleet's vessels are staffed by Naval Space Cadre officers, whose training has earned them the prestigious

6206-P Space Operations specialty designation, after they have graduated from advanced education at the Naval Postgraduate School in Monterey, California and earned a Master of Science degree in Space Systems Operations.

Secretly, UNOOSA can call on the UN's Central Security Service (UN-CSS) to coordinate with the Solar Warden space fleet.

There are 2 major secret commands under UN-CSS:

1. Space and Naval Warfare Division under MJ9 Commander John Chandeler

2. Special Operations Command.

Neither will appear on the UNOOSA's website because they are classified: http://www.oosa.unvienna.org/

From their thirty-fourth session (May12, 2014), UNOOSA's provisional agenda includes:

"Coordination of future plans and programs of common interest for cooperation and exchange of views on current activities in the practical applications of space technology and related areas."

"Contributions of space based technology for climate change adaptation and mitigation."

"Use of space-based technology for disaster risk reduction and emergency response."

"Use of spatial data and activities related to the United Nations Geographic Information Working group and the United Nations system: directions and anticipated results for the period 2014 -2015."

"Preparation of a special report on initiatives and applications for space related inter-agency cooperation."

"Means of strengthening the role of the Inter-Agency Meeting on Outer Space Activities."

"Other matters."

We certainly can guess what these other matters might be, now that we are more informed. But, officially the UNOOSA isn't ready for full disclosure to the world's people.

So, now we have a better idea on the secret space adventures of certain corporations taken over by the Nazi International. Some of these corporations are multinational and in many ways are virtually independent of any one government. Treaties like NAFTA, WTO and the TPP make these corporations even more independent of governments.

As these international corporations expand their enterprises out into space, the Earth Governments have even less control over them. This important fact needs to be addressed because while governments are supposed to serve the people, corporations serve "the bottom line" and people are of secondary concern - only to be used to profit the corporations ever more.

It doesn't take a genius to see where this is leading, and it certainly will not be a paradise on Earth or anywhere else. Neither money or economic bondage will truly make people happy. But there are still people who think fascism or the unity of government and corporation is a good plan.

Even though Hitler is now truly dead, as well as his loyal followers, the concept of fascism is still alive and well in the world and off of the world.

As I pointed out in *The Secret History of the New World Order,* the Vatican, the largest corporation on the planet, was the source of fascism in all of it's *Reichs* - from the Holy Roman Empire to the present *Fourth Reich* - that has now spread it's tentacles into outer space.

The Interplanetary Corporate Conglomerate

Many of the German corporations made tremendous profits during the war because they were using slave labor in their production facilities. And some of these corporations had interlocking agreements with U.S. corporations. Germans were using products from ITT and IBM to keep their database of citizens in their country and occupied countries while I.G. Farban worked with Standard Oil in other areas, including economically producing gasoline from coal in the Fischer-Tropsch process.

As stated earlier, many U.S. corporations were sympathetic to the ideas of fascism – not only because of fascist efficiency and profitability - but also many of these corporate CEOs were members of secret societies like Skull and Bones, the illuminati, and other Jesuit created organizations. It was the Jesuits that brought the Nazis into power in the first place.

These corporations later formed international conglomerates. These mega corporations now are virtually stateless – in that they are virtually beyond the control of any one nation. Treaties like NAFTA and TPP make this national independence even greater.

So when, the colonization of Mars and other places in the solar system was taking place by the Secret Space Programs (SSP), The Interplanetary Corporate Conglomerate (ICC) was formed and took over the colonization program. A subsidiary of the ICC was the Mars Colony Corporation.

The CEOs of these interplanetary corporations come from normal Earth based corporations, usually after they retire with their "Golden Parachutes" of astronomical in dollar value, severance packages. The ones trusted to keep secrets and willing to work off planet then join a "super circle" of corporate CEOs that provide a pool for the ICC to draw from.

According to Alfred Lambremont Webre, of exopolitics fame, Former Vice president of the U.S., Richard Cheney is a CEO of the Mars Colony Corporation and that other CEOs of the ICC are drawn from members of Skull and Bones.

The efficient fascist concepts that allowed the German and Italian "economic miracle" during the depression of the 1930s and the tremendous profits made by German companies using slave labor was not lost to the leaders of the ICC. For these reasons, a form of slave labor – more humane than that used by the Third Reich, but still slave labor, was planned for the space colonists.

The ICC and their subsidiary corporations like the Mars Colonies Corporation then hire units of the Solar Warden military space program as security forces for their colonies.

On Mars this security force was called the Mars Defense Force (MDF). The MDF was under the control of the Mars Colony Corporation.

The MDF was kept in separate areas from the colonies, with little or no contact between the colonists and the MDF. This was according to both Randy Cramer and Corey Goode.

The ICC has an entire industrial infrastructure that includes bases, stations, outposts, mining operations and facilities on Mars, various moons and spread throughout the main Asteroid Belt. They have facilities to take raw materials and turn them into usable materials to produce both complex metals and composite materials that our material sciences have not dreamt of yet.

They have separate groups of facilities that produce various types of technologies as well as each facility or plant that produces a specific component of a technology so that those working in the facilities and living in the support colonies/bases do not know exactly what they are producing. Much of the time the components are multi use and are used in crossover projects. There are facilities on Earth that operate in much the same manner that contribute to the SSP on several levels.

There are other bases on Mars that are controlled by Military/Security groups as well as some scientific outposts. These can be owned and maintained by other SSP Programs but are usually going to report to the ICC on some level since the ICC controls much of the Air Space and Security Operations on and around Mars.

Most of the security personnel that are assigned to Mars are assigned to and serve under the ICC. The military groups that will be returning to their previous organizations (SSP Groups) are kept isolated from the population and personnel who live and work on the Colonies, Bases and Industrial Facilities that they protect.

They are normally in the rather Spartan outposts that Corey Goode has described previously in other writings. He had seen a few of these outposts built from the "Ground Up". They were always quite a distance from the main underground colonies, bases and industrial facilities and spread out in a Multi-Tiered Perimeter Defensive type of system.

There are "Non-Humans" also having bases on the planet. Some of them have been there for some time and have the highly coveted larger lava tube systems that have been built out into base systems that are unimaginably huge and can securely reside millions of inhabitants.

Usually, Corey Goode served on a research spacecraft. But, there were several occasions where there were specialty equipment malfunctions that needed to be repaired immediately.

When the ICC was unable to arrange their personnel from Earth or other facilities in the Sol System to make a trip to a colony, base or industrial facility in the time needed, a request was put through to the specialists on Corey Goode's scientific research vessel. As Corey Goode would explain:

> "On these rare occasions we would fly down to the location where we would be met by 4 to 6 armed guards. We would be instructed not to make eye contact or communicate with anyone for any reason unless it was directly related to the work we were there to do.
>
> In these situations there would normally be one of our security team, an Intuitive Empath, and a scientist and two technicians along with tools and parts that may be needed.

We would be escorted directly to the location of the work. The local facilities security team would watch us very closely and then escort us directly back to our shuttle craft after the work had been completed and tested.

We were never asked if we would like a tour, invited to spend the night or stay and share a meal with the personnel or inhabitants of the facility. We did however get a chance to see some of the people. They were usually pale, unhealthy looking both physically and mentally and seemed very much like slave labor. On more than one occasion we saw four identical people carrying crates and other items around that were obviously clones.

I did notice in one colony that there was what looked like an "Art Wall" where people were hanging art work that they had drawn and painted. This was the only time I saw anything that looked like it was meant to be positive for the mental health of the inhabitants.

These were always places that we were relieved to leave. When we would visit the military outposts they were regimented, but had a completely different feel or energy about them. We were also more comfortable in some of those locations because we had actually seen and been a part of them being built at an earlier date."

So the Martian military bases were more comfortable to be around than the Mars colonies. Boy, is that a revealing statement!

Randy Cramer was trained since childhood to be a super soldier in a project called Moon Shadow along with 300 other children of which about 20% were female. Moon Shadow was

U.S. Marine Corps - Special Section operation that used MILAB technology to take and return the children from their homes. At age 17, Randy received some special training and then was ready for the next step.

Randy Cramer says that on November 17, 1987 at 2:30 A.M. in the morning, a familiar wormhole materialized in his bedroom and took him to an underground base. There, he was placed on-board a TR-3B craft (described in *Secret Science and the Secret Space Program*) and flown to the Lunar Operations Command on the moon. At the LOC Randy Cramer was taken to an office and was signed up for his 20 year tour of duty in the Earth Defense Force.

The Earth Defense Force (EDF) is a U.N. Unacknowledged Special Access program, which recruits it's personnel from many military services globally.

After signing the contract, Randy Cramer was taken to a huge hangar and observed a huge Triangular spacecraft about 150 feet long and 4 stories high. The craft could carry over 2,000 passengers in his estimate. He boarded the spacecraft with other EDF recruits, and many empty seats, and was soon high above the Moon.

Then the Craft stopped and the ceiling suddenly became transparent, acting like a huge window into space. In the middle of the star filled blackness of space was the blue planet Earth. A voice over the intercom informed the passengers to take a good look at the planet they were defending - it would be a long time before they would see it again.

Then the spacecraft entered a natural wormhole or portal and navigated to Mars in about 15 Minutes. When the Spacecraft landed on Mars the recruits were ordered to leave the craft through an opened door. He was surprised that he could breathe the Martian air, which was cool and thin, like on a high mountain. Then, he was directed towards a nearby building where he would be briefed and given his orders.

After reaching the Building , he was informed that this was the *Aries Prime* base, the primary headquarters of the Mars Colonies, which had been built by the Mars Colony Corporation (MCC). He then was informed that now he was a member to the Mars Defense Force, a subsidiary of the Earth Defense Force. Since the MCC had hired the EDF as a private contractor the MDF was under the corporate control of the MCC.

Then he was loaded aboard a shuttle and taken to forward station Zebra in the Northern sector of the colonies. He would live and fight at, and around, station Zebra for the next 17 years. These military posts that defended the colonies were kept quite separate from the actual colonies and contact between the colonists and their military guardians was strictly limited.

Randy Cramer knew of 5 separate colonies on Mars. Corey Goode states that the asteroid belt is being mined and there are colonies and military bases on the moons of the larger planets as well as satellite production facilities.

Members of the MDF can return to Earth after their 20 year tour of duty. The colonists are there forever.

Many years after retirement from the Solar Warden fleet, Corey Goode was asked to join the SSP Alliance.

The SSP Alliance was formed by members of the Solar Warden fleet who took seriously their oath to uphold the constitution for the United States. They decided to take action to defend the freedom of humankind in the solar system, and started to make war on the fascist ICC and their outposts.

After a disaster which wiped out an entire Mars colony, negotiations were entered into between the ICC and the SSP Alliance.

The ICC claimed that their colonies had no slaves. The SSP Alliance challenged that assertion by requesting an inspection of any Mars colony of their choice. After long negociations it was decided that not only could the SSP Alliance inspect a Mars colony of their choosing but that they also could select a willing family there to be taken to a non ICC controlled place for discussions without fear of reprisal.

On one recent occasion, Corey Goode was asked to join in an inspection of a Mars Colony to observe the working conditions there by another SSP Alliance member, who for his safety wishes to remain unnamed, and who will use the alias Gonzales.

Corey Goode explains what happened when they arrived at the Mars Colony they choose to visit in the southern hemisphere of Mars.

In this tour of a colony,which for them was basically enemy territory, their safety was guaranteed by an ICC representative that accompanied them. As they approached the base commander, Goode explains what happened:

> "The ICC representative then told us what the subject of that conversation was going to be. He was carefully watching our reactions as he gave us a summary of what to expect from the base commander.

He stated that the people at this facility were here for generations and that they were under the impression for decades now that the earth had been through some sort of cataclysm and was no longer inhabitable.

He said that we are being asked to not throw off the social dynamics of the facility by revealing that this was not the case or that any of us were from the surface of Earth. He then turned around and did the same half jog back towards the facility security team while we looked at each other wide-eyed regarding what he had just said. Gonzales said "You heard the man" let's not have any incidents here if we can avoid it.

We then filed out of the SSP Alliance vessel and headed towards the security checkpoint. We were stopped immediately when a very cold looking security personnel looked at our security contingent and said "Absolutely NO weapons past this point!" We stopped and Gonzales looked at the ICC representative and stated, "You know that is a deal breaker, I'd hate for things to end before they got started here wouldn't you?"

The ICC representative walked past the checkpoint and down the hall a ways and spoke to the base commander. The base commander was visibly agitated from the beginning and this did not make him any happier.

The base commander called to his security and waved us through. This was another obstacle overcome by Gonzales, so far so good.

As we entered, the base commander gave us the storyline about the people not knowing that the earth was still thriving and that this wasn't anything to do with any slavery theories being promoted by the SSP Alliance but was a complex "social experiment". It was stressed that we needed to be careful not to contaminate a multi-decades long experiment that will help humanity.

I looked at Gonzales and he rolled his eyes at me as this was being said. The base commander then stated that the "main hall" was being setup for the conference and that we would be taken on a tour of the industrial plant first that was 8 kilometers away via an underground train. He said after the conference we would then tour the colony and meet the people, see their living conditions and ask a family if they would be willing to leave with us.

We were walked through an area of the colony that had obviously been cleared of people for us to walk through. The area was a ghost town and that much usable space is not wasted in these facilities. We made our way to another area that was also cut out of the rocks and unfinished. There was a single monorail track going in one side and out the other of a tube that was glass-like but not smooth or polished. It was ribbed or bubbled it seemed. It was possibly vitrified rock? A very thin train then arrived and we were told to load up.

The way we entered caused us to be seated in a strange order. The seats were two by two and facing each other. I ended up being several people down from Gonzales and had one of our security personnel on either side of me. There was a solid row of the facilities security personnel across from me. I noticed one of them kept whispering to the other and finally asked me as we were traveling down the tunnel "Where are you from", I responded without thinking "Texas" and received a very shocked look on the 3 faces that I could see in front of me. I knew immediately that I had already screwed up.

The facility security began to immediately talk among themselves and I could feel the eyes of our security team IEs [Intuitive Empaths] on me and within a few seconds could see an outstretched neck and a head turned my way from the general area that Gonzales was sitting. Yea, I screwed up big time. I didn't want to look at anyone else from my team and just kind of looked out the window behind the facility security team and watched the blurry wall until we arrived at the industrial plant.

As we disembarked from the train, the murmurs were growing louder among the facility security personnel. I then came face-to-face with Gonzales who had a smart ass look on his face and a smirk. I looked at him and said "Yea, I screwed up". I then told him what happened and he said that he had heard them discussing the way he and I smelled and the fact that I was sunburned and that I didn't look like someone that has been on a colony or stationed on a vessel.

He went on to say that people in these closed environments can tell each other apart. When someone like us comes in smelling like hair gel, deodorant and coconut scented tanning oils and aloe vera sunburn cream we reek of Earth and smell very alien and out of place to them. These people all use the exact same toiletry products and even the smells of the foods that come out of the pores in our skin are a dead giveaway.

He said it certainly doesn't help when one of us says we

are from "Texas". He said he was brainstorming on coming up with an off the books facility or other vessels or locations that he could say were code named Texas but he thought the damage was done. This proved to be correct halfway through our tour of the industrial complex. It was obvious that the talk was spreading through this facility security team very quickly and the ICC representative was very aware of it. He was making sure we saw him glancing back and forth between the people chattering and our group with a disapproving look on his face.

We were all walking in a close group as one of the residents was explaining what each of the robotic machines did and along the conveyor systems what each of the people did at the work stations in the process of producing the various shaped panels that were optical and neurological relays and displays. We were seeing a version that was a slightly curved panel of one size but they stated down the tunnel system there were other rooms where the same panels were produced in various configurations and for various other types of biological neurological interfaces (non-human customers).

It was at about this time that another monorail train arrived with many more security personnel. They told us to halt and they separated the security team that was with us, disarmed them and escorted them to the trains and left. A new security team was assigned to us and the ICC representative that obviously had an "earwig" (communication device in his ear) told us that we were not to communicate with the security team unless it was to do with something security related. They then brought over a metal cylinder that had a spray atomizer on it and told us to spray it on our bodies and rub it into our faces and hair. It was an odor neutralizer and it didn't have a scent of its own.

I was worried that it had something in it but Gonzalez said its okay and I unzipped my jumpsuit and pulled it down to rub the solution into my arms, shoulders, neck, hair and face. We then ended the tour early because the ICC representative stated that the other ICC leadership had arrived and were setting up for their presentation in the "main hall".

We waited for the monorail train to arrive back, piled back on it and headed back to the colony. On the way back one of our security/IE personnel leaned over to Gonzales and I heard him say "Something's not right". Gonzales nodded to him and said "keep me informed."

"IE's" get false hits often and it's good to have at least 3 present to triangulate any threats. When we would point out a possible feeling we were always told to stay on the scent and report any new hits.

When we arrived at the colony there was a completely different energy. There were people everywhere that all appeared between the ages of 8 years of age and about 60 (at the oldest) bustling around in one piece suits that were obviously their "Sunday best". They were also all trying to look like they were going about their business while trying to sneak a peek at the new arrivals and seemed to be trying to make eye contact. It seemed that they had been told that we would be touring the facility and asking one family to leave with us to give a report about the facility and how it is ran and what it's like to live there.

We were brought into the "main hall" that looked like an area that people are brought in for daily propaganda and there were a large number of ICC leadership members present who were bustling about as well and it was difficult to count how many were there. They sat us down and put on a "Dog and Pony Show" on a large "smart-glass screen" that showed all sorts of technologies that they produce, what they procure in trade for those technologies and stated that they had ongoing trade agreements with almost 900 civilizations and did occasional trade with far more than that.

They showed all sorts of spacecraft and spacecraft components that some groups integrate into their own technologies and also discussed the exopolitical agreements they had made with groups that pass through our sector on a regular basis using the nearby natural portal systems that are a part of the "cosmic web". After this presentation we were taken on a tour of the colony. People were very eager to please and show us their dwellings that were the size of my dining room and housed a family of 4. They had very little in the form of physical possessions and there seemed to be a cast system that ironically was similar to the movie Divergent but on more of a micro level.

We had gotten to the end of the tour where we had seen their environmental control system and their laundry and recycling system (everything is recycled) and community centers.

It was now time for Gonzales to give the ICC representative the choice of the family we were to take back with us. He gave the number of a certain family's dwelling that was an alpha numeric number outside their door and 15 minutes later a man, woman, teenage son and pre-teen daughter showed up with a small bag in each of their hands. They seemed kind of stoic and nervous. This was to be expected but they seemed off to me. I didn't say anything at the time. One of the "IE"/Security Personnel leaned over to Gonzales and whispered to him and Gonzales ordered us all to our vessel.

When the door was closed Gonzales turned to the people and told them they were safe and that he would not betray their good faith. He said that "We know that there is another member of your family that is not present", the father said "How could you know that?" Gonzales motioned to our security personnel and said that "our people have abilities". The family then clammed up and would not talk. Gonzales became upset and said he would straighten this out and he and the two security personnel assigned to him left the vessel.

We sat in uncomfortable silence for about 10 minutes and then all 4 of the crew came out and said "We have over a dozen of the facilities security personnel coming out fast with weapons high towards the craft". They asked me "What should we do?" I asked if they saw Gonzales and they said no, he was not out there. I told them to open the doors and to stand down. The crew enacted security measures that wiped the systems of information that would be helpful to the enemy. When the doors to our vessel opened the facility security personnel then entered and disarmed our security personnel and took us into custody. I had that sinking feeling in the pit of my stomach knowing that something had gone wrong and wondering if Gonzales and the others were okay.

The facility security team walked us in, straight past the base commander that was in a very heated conversation with the ICC representative. Something had gone very wrong and his ego was bruised or his authority was challenged in front of his men. He was extremely upset and was no longer listening to the ICC representative who outranked him.

We were walked to a wall that suddenly had a double door where there was none before. They opened it and we were then walked into a detention facility that was conspicuously absent on the previous tour. As we walked back through the rows of cells we saw quite a lot of people in various psychological stages of psychological distress locked up. When we reached the back cell where Gonzales and the two other security personnel were detained I was relieved to see them alive. We were all locked up and left without a word spoken to us in the same cell.

Gonzales said that the base commander was a tyrant and a total megalomaniac who was not used to people challenging him diplomatically or otherwise on his base. He said that the recent attack that the rogue SSP Alliance Forces had done on Mars had come up in the argument and was obviously a point of contention that the base commander used to toss him in the brig. I asked what was going to happen and he said that from the way the base commander was talking it didn't sound good, which then left all of us thinking the worse.

It wasn't a minute later that we saw our blue/indigo orb friends zipping through the walls of the cell, one for each of us. They danced around for a couple seconds and the other SSP personnel who had never traveled this way backed against the wall.

Gonzales explained how the transportation works and then we each followed the SOP and were soon back at the LOC room where I was transported at the beginning of this journey. Upon our unexpected arrival there was an intrusion alarm going off at the LOC and armed security came into the room very quickly. Gonzales told the SSP Alliance personnel to report to their units for debriefs. They left the room with the armed security teams and the door was closed leaving Gonzales and myself alone in the room.

Gonzales then said "We lost a vessel but didn't leave any team members behind, that's something". He stated that we received quite a lot of good intelligence which was the real goal of this mission. He stated that the ICC's goal was propaganda and that since the recent reports that were released to the public about the slave trade and labor going on by the ICC that they have been extremely upset and worked-up about information being public that was never supposed to be so. He told me to continue doing what I am doing and that he didn't know where I was going from here. He said he didn't know if I was going to visit Raw-Tear-Eir or be taken back home. We said our goodbye's and he left to begin his debrief process." (9)

Corey Goode's story gives a good idea of what a Mars colony is like. It certainly doesn't seem like human rights are given much consideration by the Interplanetary Corporate Conglomerate there.

The Coming of the Sphere Beings

Sometime before 2012, as the Sun started its trajectory from the upper part of the galactic disk to the lower part and entered a region of more intense vibrational energy from the galactic core, many prophecies of Earth upheavals and destruction were taking place as you probably well remember. There was evidence from satellite data that *all* the planets of the Solar system were heating up – not just Earth.

It seems that other extra terrestrials and extra dimensional civilizations were also concerned about effects of this higher vibrational energy which was coming in, in the form of Tsunami waves of energy that would ebb and flow, would have on the solar system and the life forms residing thereon.

So a group, that Corey Goode describes as the Sphere Being Alliance decided to take action to minimize damage to planets within the Solar system. They manifested thousands of huge planet size spheres that are cloaked , equally spaced, throughout the solar system and neighboring solar systems in the local star cluster. These spheres are able to stabilize the energy waves impacting these star systems and lessen the planetary upheavals of these star systems.

It turns out that the presently accepted theories on what the sun is are totally wrong. The sun is an electrical powered – not atomic fusion powered - energy source. This electric power is transmitted via a cosmic web, with each star being a extra dimensional portal connected with each planet, each local star in the local star cluster and to the center of the galaxy. Likewise each galactic center is connected to other galaxies in the local galaxy cluster. This is the cosmic web.

The surface temperature of the Sun, which is a plasma of electrically charged particles, is much hotter than the interior of the Sun. The sunspots are huge tornadoes within this plasma which open up to the deeper portions of the Sun's photosphere which are much cooler, and is why they appear darker. If the sun was fusion powered, the interior would be much hotter than the photosphere - not the other way around!

Much of this information was revealed by Corey Goode from information he received while working on a Solar Warden research vessel, but some Earth scientists have also figured out parts of this information. For example, check out: http://www.electricuniverse.info/Introduction

I have seen videos on the web showing , planet sized, spheres close to the Sun. Whether this was one of the Sphere Being Alliance operations, I can't say. But it could be, and it lends more credibility to Cory Goode's testimony of this alliance: https://www.youtube.com/watch?v=KucBPnoKNpg

Here is more evidence of planet sized orbs near the sun: https://www.youtube.com/watch?v=xmEU-x2IEOk

In any case, representatives of this Sphere being Alliance, called the Blue Avians, contacted members of the Secret Space

Program(SSP) Alliance .

This SSP Alliance was gradually formed to counter the present fascist cabal that rules the earth through; secret societies, lies, and the criminal banking system, along with the Draco reptilian alien overlords and bring an end to the slave colonies on the planets around our Sun.

Most members of the SSP Alliance were from the Solar Warden space program, who took their oath to uphold the U.S. Constitution seriously. But before long, there were defectors from the Interplanetary Corporate Conglomerate (ICC) and other secret space programs, who also felt qualms about the agendas of their employers, that also joined the SSP Alliance.

This defection speeded up after the Sphere Being Alliance placed an energy field blockade around the solar system, around 2012, which prevented entering or leaving it. Even portal travel, which was becoming more unstable and dangerous with the incoming higher energy waves, was blocked.

Persons in the Global Galactic League of Nations working outside the solar system were stranded there. And members of the Draco Alliance that were on Earth or elsewhere in the solar system were prevented from returning to their home worlds.

Often these defectors would bring the more advanced technology of their programs to the SSP Alliance people, who were using older Solar Worden technology. Also, the Sphere Being Alliance, who were very much opposed to violence, contributed purely defensive technology to the SSP Alliance.

The Sphere Being Alliance contacted the SSP Alliance because they had made a conscious decision to counter the extra terrestrial "Gods" that the cabal worshiped, and the evil cabal controlling the Earth and the ICC colonies of the Solar system. The Sphere Being Alliance wished to free the people of Earth to evolve to greater spiritual awareness.

It was a member of this SSP Alliance that contacted Cory Goode, years after he had retired from Solar Warden, to become a member of this alliance. He was selected because of his abilities as an intuitive empath. They wanted him to attend conferences between different ET and Earth based groups and use his abilities to detect falsehood or possible danger from representatives of these other groups.

When Corey Goode was a child, he was observed to have physic ability. The Military Abductions (MILABS) would train him to enhance these abilities. Latent physic ability can be greatly enhanced in childhood between five years of age and puberty before the mental programming of the physical world becomes too fixed in the person's mind.

MILABS is the experience of abductions, not by aliens - but by human military personnel. And the experience of telepathic conversations with human military personnel. Military abductions are carried out using alien abduction technology.

This technology can control time so that an abduction of a person can occur, hours of training take place, and the person returned at close to the same time he or she was abducted.

Teleportation is also used so the person could be taken right out of bed at night, right through walls or ceiling and returned the same way.

In this manner the MILABS abduction of children can be hidden from their parents. If the child talks about their training experience to their parents – the parents think that their child has an overactive imagination.

Abductees were often planned for before birth and the children have alien DNA which was placed in their parents usually through similar abductions.. Abductees with alien DNA have special alien mental and psychic capabilities such as telepathy.

Usually, the military wants to use select MILAB persons in military operations. This has been going on for several generations with the military - and longer with the aliens, using their own technology.

After Corey Goode was trained in this manner and later became of age, he was taken to the Lunar Operations Command on the Moon, and signed up for a 20 year hitch in Solar Warden.

Much later, when he had been retired from the Solar Warden program for a number of years, he was asked to join the SSP Alliance and accepted. That is when he came into contact with the Blue Avian representatives of the Sphere Being Alliance. They are called the Blue Avians because they are covered with blue feathers and have a very short beak like mouth.

The Blue Avians are extra dimensional beings and possess amazing abilities themselves. Corey Goode was assigned to work with a Blue Avian named Tear-Eir at the conferences.

Besides the Sphere Being Alliance, there are other Ancient breakaway civilizations on the earth with their own secret space programs. Most of these Ancient Breakaway Groups are very connected to the "Universal Mind" (Akashic Records), The Law of One philosophies and the true Physics Nature of the universe as well as the way Densities and the completely different topic of Dimensions and Realities Work.

They were very quiet and seemed to "Sit Back" and observe the current era SSP's and Secret Earth Governments as have the "3rd - 4th and 4th - 5th Density Off-World Groups" who also have been figuring things out fairly quickly.

Some of these "Off-World" groups (Draco/Orion & Others) have shown recently that they are willing to sacrifice some of their own lower caste as well as ALL of their "Elite Human Following On Earth" in exchange for clemency and to be allowed to leave Earth and the "Outer Barrier" of the Sol System.

They have made these "Direct Offers" in the forms of "Petitions" to the Sphere Being Alliance and have been denied! Some in the "SSP Alliance" (and recent "Defectors" who have joined) thought this would be a "Win/Win" situation and were not at all happy while "Others" in the SSP Alliance were very much behind the decision.

The "SSP Alliance Rank and File Soldiers" had some problems with the information after being "Fully Briefed" on the Sphere Beings and the "Blue Avians". Most if not all of these personnel have been drafted through the MILAB Programs. One of the larger MILAB Programs in particular was Project Blue Bird.

The personnel who had come from "Project Bluebird" were having a very difficult time dealing with the existence of the "Blue Avians" after the briefing. Some of them were "Triggered" and there were some very unpleasant incidents of which Corey Goode ended up directly intervening in at one of the locations.

These soldiers, many of who had deep seated subconscious

trauma from Project Blue Bird, at first believed that someone was trying another brainwashing job on them. Finally, Corey Goode was able to convince these soldiers that the Blue Avians were the "good guys."

There are a few people that are working directly for "Cabal/ Illuminati Disinformation Cells" down here on Earth and on the Internet that have started "Blogs" and other followings on Forums stating that the "Sphere Being Alliance" and the "Blue Avians" are a giant "PSYOP" and using very clever mixed information they have stolen and meshed together from "The Hidden Hand", "The Law Of One" and the "SETH" materials. Always use discernment and listen to your heart in these cases.

Sadly, advanced technology and spiritual advancement do not go hand in hand. Several of these "Ancient Breakaway Civilizations" have been very dishonest and have presented themselves as "Gods" and "Aliens" from other star systems here to "Help Humanity". They have taken advantage of people of Earth for many thousands of years and worked with the most corrupt "Off World Groups" when it suited their agenda and with the "Secret Earth Governments and their Syndicates" up until recent times.There seems to have been a falling out with some of these groups and the Secret Earth Governments because they are trying to trick others into exposing the Ancient Breakaway Groups who are pretending to be "Aliens" for what they really are.

The Secret Earth Governments and their Syndicates are also in the process of being betrayed by their "Custodian Gods" or "Off World Allies" who are trying to make deals to sacrifice them for the freedom of these "Off world Groups". These "Opposition SSP's" and the various Secret Earth Government Syndicates are also turning on each other at this time.

They are barely holding things together at all. The "SSP Alliance" has its own problems with cohesion as well. It is quite a mess with all groups both "Above" and "Below". This could be the breaking points that Humanity needs right now.

The Blue Avians stated that things would get a lot tougher before they improve not only because of the nature of "Full Disclosure" and the revelation of the Crimes against Humanity, but also because humanity needs to have some karmic releases and experience some hard lessons that will stick in our genetic memories that will prevent us from repeating these historical cycles once we are free from manipulation and control by the "Custodian Alien Races/Gods" and their Human Control Systems, the "Secret Earth Government's, Their Syndicates and the "Banking Economic Bondage System" that have almost always been in place.

We will then be responsible for our own futures. I cannot think of a better foundation to start that future off on that that of their Message. To become "More Loving", To Daily work on becoming "More Service To Others", To practice "Forgiveness" (To release karma) and to focus on "Raising our Vibrations and Consciousness".

The message of the Sphere Beings is quite identical to the message of Jesus Christ and other world teachers of high spiritual evolvement. The people of the Earth need to become more loving and of service to their fellow human beings.

Persons that are primarily self serving and that inflict harm on others will not be able to continue living on this planet as it ascends into higher vibrational energies. They will be removed to other places of lower vibration to live with others like themselves and work out their karma.

The Sphere Beings manifest themselves as blue violet spheres varying in size from golf ball to beach ball size. These spheres seem to suddenly manifest in front of Cory Goode and slowly approach him. When he telepathically communicates that he is ready, the sphere expands enough to enclose him and quickly transport him right through walls and space to the SSP Alliance headquarters or elsewhere.

It was this ability that the Sphere Beings used to free Corey Goode and his companion from the Mars colony prison to continue with their mission towards full disclosure and true freedom for the Earth's and solar system's people.

One of the effects of the higher density energy now engulfing our planet has on people is that it intensifies their energy. If these people are on a path of love fairness and peace and seeking to be of service to others – they will become more that way. On the other hand, if they are on the path of causing harm to others for their own selfish gain, that feature of their self will intensify until they karmically cause their own self destruction.

This action is due to the cosmic principle that like attracts like, as summed up in the saying "birds of a feather flock together."

The selfish and destructive people will attract other selfish and destructive people around themselves and they will end up destroying each other.

And, the loving people, serving others, will end up creating a paradise on Earth. This is encapsulated in the Bible, "Do unto others as ye would have them do unto thyself." and "The meek shall inherit the earth."

Some might dismiss the above information as "hippy or new

age nonsense". But believe me - a person that has been around this plane for three quarters of a century and experienced a lot - I have seen this principle happen many times around myself and others.

Scalar Physics

This chapter is included for the person with math and physics training that would be interested in the theory of electrogravitation and portal physics and what mistakes are in the currently accepted theories. The more casual reader could skip this chapter without losing the main context of this book.

Scalars are different than vectors. Vectors have both a magnitude of some physical quantity and a direction in 3 dimensional space. Scalars on the other hand, have the magnitude of some physical quantity, but have no direction in space. Examples of scalars are temperature, pressure and voltage. Examples of vectors are force, velocity, and magnetic fields.

Performing a mathematical operation known as a gradient, ∇V, which takes the partial derivative of the scalar and how it changes with respect to change in each spatial coordinate, produces a vector in the direction of greatest change.

For example, if in a space where the greatest voltage difference is between point P1 and point P2 and if at point P1, the voltage is 30 volts and at point P2, the voltage is 10 volts and the distance between P1 and P2 is 1 meter, the gradient of the voltage creates the electric field vector, $\mathbf{E} = -\nabla V$, pointing from P1 to P2. with a magnitude of 20 Volts per meter.

On the other hand, another operation called the divergence , $\nabla \cdot \mathbf{E}$ - which takes the partial derivative of each spatial coordinate component of the Vector $\mathbf{E}$ with respect to that spatial coordinate - will produce a scalar value of charge density divided by the constant of electrical permittivity of free space. Electric field lines radially diverge from a charge.

The curl, $\nabla \times \mathbf{A}$, of a vector is more involved , but essentially creates a vector space that curls around the original vector much as the magnetic field, $\mathbf{B}$ curls around a electric current, $\mathbf{I}$.

Scalar Physics also unifies gravity and electromagnetism. The forces of magnetism, electricity, and gravity are distortions of a single primordial field that permeates the universe and comprises the fabric of existence. Vorticity in this field gives rise to magnetic fields. Dynamic undulations give rise to electric fields. Compression or divergence gives rise to gravitational fields.

When put into mathematical form, these relations reveal how electric and magnetic fields can be arranged to produce artificial gravity and many other exotic phenomena such as time distortion and the opening of portals into other dimensions.

Scalar physics is the science of reality's hidden understructure. The electric, magnetic, and gravitational force fields are only the surface layer. These forces arise from deeper fields known as *potentials*, which themselves arise from a primordial *superpotential,* which some scientist consider the aether.

The scalar superpotential is the substrate of physicality, the aether permeating and underlying the universe, from which all matter and force fields derive.

It is a scalar field, meaning each point in that field has one value associated with it. This value is the degree of magnetic flux at that point, whose unit is the Weber. This is not the magnetic force field we all know, composed of vectors whose units are Wb/m2, but a magnetic flux field of scalar values whose unit of measure is simply Wb. Also 1Wb = 1Newton Meter/Ampere.

Its symbol is X (Greek letter chi). Scalar superpotential may be written as: $X = X(x; y; z; t)$, an equation assigning a flux value to each coordinate in spacetime.

superpotential → potentials → force fields

Force fields derive from specific distortions or undulations in potentials:

- Gradient in the superpotential [X] → magnetic vector potential [A]

- Change over time in the superpotential [X] → electric scalar potential [V]

- Vorticity in the magnetic vector potential [A] → magnetic field [B]

- Gradient in the scalar electric potential [V] → electric field [E]

- Gradient in the gravitational potential [P] → gravity field [G]

In conventional electromagnetic theory, the Curl of the Magnetic

Vector Potential, **A** is the magnetic Field, **B**. And the negative derivative with respect to time of **A** is the Electric Field, **E**.

Maxwell called the Magnetic Vector Potential, **A**, - which points in the direction of electric current in the space surrounding a current carrying wire - electromagnetic momentum.

Now, if a gradient of the superpotential, X creates the Magnetic Vector potential **A,** the negative partial derivative of **A** with respect to time creates the electric field, **E** and the curl of **A** creates the magnetic field, **B** - what does the divergence of **A** create?

What is not yet officially realized is that the divergence of the Magnetic Vector Potential, **A** is proportional to the gravitational potential, **P**. And, the gravitational force field, **g** is equal to the negative gradient of the gravitational potential **P**. (10)

$$P = k \; \nabla \cdot A$$

$$g = - \; \nabla P = - k \nabla (\nabla \cdot A)$$

Here k is a constant of proportionality.

The baseline value of a potential cannot be detected by standard instruments, neither will a change in this value always cause a corresponding change in the electric or magnetic field. What science cannot measure absolutely it sets arbitrarily to whatever is most convenient. This is called setting the gauge. The ability to choose the gauge freely is called gauge freedom.

The Coulomb Gauge sets the divergence of the Magnetic Vector Potential to zero:

$$\nabla \cdot A = 0$$

This is the arbitrary setting of the divergence of the vector potential to zero, which unbeknownst to modern science is the case where there is no gravitational potential - an impossibility on the Earth's surface.

Another gauge comes in response to the question "What changes can we make to the potentials comprising an electric field without disturbing that field?" This is known as the Lorentz Gauge:

$$A = (1/c^2)(\partial \varphi / \partial t)$$

Here φ represents the scalar potential, or volts.

The Coulomb and Lorentz gauges, while conveniently set to keep magnetic and electric fields isolated from unintended influences of the potentials, simply state the unique conditions where that isolation exists.

The Coulomb gauge sets the divergence of A to zero, signifying the one condition where there is no gravitational potential and where physics may continue undisturbed by that extremely useful possibility.

Likewise, the Lorentz gauge sets the one condition where potentials may change in mutually canceling ways without affecting the measurable force fields.

So by employing these gauges, science and engineering unwittingly limit themselves in their experiments and technology to only those applications where there are no electrogravitic or gravitomagnetic phenomena. And then they claim there is no proof of such phenomena, failing to see that their own arbitrary choice of gauges quarantines them from witnessing such proof in the first place.

Regauging relative values to zero, or pairing them in mutually canceling opposites, is how exotic phenomena are swept under the rug. This sleight of hand is the fundamental reason why humans today are facing an environmental crisis from the use of primitive energy and transportation technologies.

The general electromagnetic wave equation in terms of the Magnetic Vector Potential, **A** is:

$$\nabla^2 A = (1/c^2)(\partial A^2/ \partial t^2)$$

Here $\nabla^2 A$ is the Laplacian of **A** and $(\partial A^2/ \partial t^2)$ is the second partial derivative with respect to time of **A.**

Both the electric and magnetic wave equations can be derived from this. What we visualize as electric and magnetic fields fluctuating into each other while propagating through space as part of an EM wave, may in actuality be a single magnetic vector potential wave.

In fact, the primacy of the vector potential demands that it be more "real" than either, with the electric and magnetic components just being derived abstractions.

This is important because one of the arguments that there is no ether was made on the basis that magnetic and electric fields can sustain each other while traveling through a vacuum, but if the real wave is a single vector potential wave without a supporting partner, then there must be a medium of propagation - the medium of ether which may identically be the Superpotential.

The general wave equation for **A** can be rewritten using the vector identity:

$$\nabla^2 A = \nabla(\nabla \cdot A) - \nabla \times (\nabla \times A)$$

Therefore:

$$\nabla(\nabla \cdot A) - \nabla \times (\nabla \times A) = (1/c^2)(\partial A^2/\partial t^2)$$

Here $\partial A^2/\partial t^2$ is the second partial derivative of **A** with respect to time and c^2 is the speed of light squared . Notice that the left side has two spatial distortion components, the first a gradient in divergence and the second a curl of curl (or curl of magnetic field). If we choose the Coulomb Gauge where $\nabla \cdot A = 0$ then:

$$\nabla \times (\nabla \times A) = -(1/c^2)(\partial A^2/\partial t^2)$$

Or in terms of **B** and **E**:

$$\nabla \times B = -(1/c^2)(\partial E/\partial t)$$

This also happens to be Maxwell's fourth equation per Oliver Heaviside's reformulation, which implicitly contains the Coulomb gauge. As can clearly be seen, Heaviside eliminated $\nabla \cdot A$ from Maxwell's original work, eliminating the gravitational aspects of electromagnetism.

So, let's look at the case where **B** = 0 and $\nabla \cdot A$ is non zero:

$$\nabla(\nabla \cdot A) = (1/c^2)(\partial A^2/\partial t^2)$$

This is the longitudinal wave equation for **A**. The spatial component is entirely gravitational. Since $g = -\nabla P = k\nabla(\nabla \cdot A)$:

$$g = k(1/c^2)(\partial A^2/\partial t^2)$$

Where k is a constant of proportionality, **g** is the gravitational field and /c^2 is the speed of light squared.

This equation implies that a nonlinear time change in **A** produces a gravitational force. Since **A** is proportional to the electric current generating it, a nonlinear current change will produce a gravitational force as well.

Hence the phenomena of exploding wires and buckling rail-guns, whereby nonlinear current pulses produce longitudinal gravitational forces that snap or warp the metal.

In terms of **E** we finally arrive upon the electro gravitational field/wave equation.

Scince E = - ∂A/ ∂t
g = -(k/c^2)(∂E/ ∂t)

The value of k can be computed from the above equation knowing that the electric force, **E** is 2.27 X 10 ^39 stronger than the gravitational force, **g**. and **c^2** = 9x10^16 meters squared per second squared. This also means that it is entirely practical to produce gravity or anti-gravity electromagnetically.

So, a changing electric field produces a gravitational field. But doesn't Maxwell's fourth equation say it produces a curled magnetic field? Well, that depends on the case. Depending on the geometry of the electric field, it can give rise to **B** or **g** or a mixture of the two. A

long thin metal antenna will radiate mostly transverse electromagnetic waves.

Meanwhile, a flat metal plate or sphere suppresses the magnetic field and radiates primarily longitudinal electro gravitational waves. Nikola Tesla used spherical antennas which could only transmit longitudinal electric waves as the electric vectors are always normal to the conducting surface and **E** would radially be in the direction (or against) of wave travel. A curled magnetic field would be impossible to create with this geometry which would only transmit longitudinal electric waves.

Unlike electromagnetic waves, these electro gravitational waves can penetrate matter - even conducting matter - just as gravity does. Classified communication technology uses these longitudinal waves for undersea communications between submarines and undersea bases.

According to General Relativity, the equation for time dilation (slowing of time due to presence of gravity) as a function of distance from an attracting mass is as follows:

$$t = t0/ (1-2Gm/rc^2)^{\frac{1}{2}}$$

where m is mass, r is radius from the center of mass, c^2 is the velocity of light squared, t is the dilated time, to is the original time, and G is the gravitational constant. The gravitational potential P as a function mass, m, and radius, r, is:

$$P = - Gm/r$$

Therefore the time dilation equation may be rewritten in terms of the gravitational potential P, and thus, since $P = k(\nabla \cdot \mathbf{A})$ as a function of the magnetic vector potential:

$$t = t0/ (1+2\mathbf{P}/c^2)^{\frac{1}{2}} = to / (1+2k\nabla \cdot \mathbf{A}/c^2)^{\frac{1}{2}}$$

This implies several things. First, it says that gravity is a time gradient; as you get closer to an attracting mass like a planet or star, time slows down for you relative to the rest of the universe because the gravitational potential is becoming more intensely negative.

Second it implies that a diverging magnetic vector potential (like that inside a hollow sphere given a voltage signal) will affect the time rate, speeding it up if the divergence is positive and slowing it down if negative.

And such a field can be created artificially, thus the key to using electric, magnetic, or electromagnetic fields to alter the time rate is to simply create a divergence in the magnetic vector potential.

Third, by setting the equation to zero, the equations show that if the gravitational potential P, is lower than $-C^2/2$, time slows to a stop. This condition may be found at the event horizon of a black hole. Beyond this point, time becomes imaginary relative to the rest of the universe and any matter inhabiting such a zone is severed from the space-time continuum and ejected into imaginary space.

According to Relativity, that would set the portal condition for the gravitational potential:

$P < -C^2/2$

Or in terms of **A:**

$k\nabla \cdot \mathbf{A} < -C^2/2$

This means one can tear open a portal into other dimensions, establishing a singularity like a black hole minus the destructive gravitational forces because this can be done electromagnetically!

Now let us turn our attention to scalar electromagnetic waves, as opposed to transverse or longitudinal electromagnetic waves.

Scalar electromagnetic waves are entirely different than longitudinal electromagnetic waves in that they have no direction in three dimensional space. According to Thomas Bearden, when 2 diametrically opposed (180 degrees out of phase) electromagnetic waves combine, their vector components cancel, leaving only a scalar electromagnetic pressure component wave – not a zero wave as predicted by classical electromagnetism.

This scalar wave oscillates in the time dimension. Since the same instant of time exists everywhere in the cosmos, as experimentally determined by Kozyrev and Shipov, scalar waves travel instantaneously to any part of the universe. This concept violates Einstein's theory of relativity on several points.

Torsion waves and scalar waves have a lot in common. Both are transmitted instantaneously with no speed of light limitation. Both types of waves penetrate the electron shells of matter and directly affect the atomic nucleus. In fact Kozyrev and other researchers believe that they are identically the same.

Working at the Pulkovo Observatory, Kozyrev discovered that the Sun was hollow and that the center was rather cool.

He theorized that the sun was powered by scalar energy which was responsible for solar phenomena, like sunspots. He observed that scalar energy was a spiraling energy that itself was a flow of time which acted on the aether to power the stars.

Scalar energy is now recognized as a spiraling double helix of phase conjugated electromagnetic energy by some researchers. This phase conjugation and spinning probably occurs when oppositely directed photons interact because these photons have spin.

Kozyrev also experimentally proved that the transmission speed of scalar waves was at least 9 powers of 10 greater than the speed of light if not instantaneous. This observation served to establish that scalar energy preexists as a connection between the stars and the earth and this preexisting connection is responsible for the instantaneous communication over these vast distances.

This demonstrates that the universe is a hologram where each part contains all of the rest of the universe and all parts and times, in some way, are intimately connected. And herein lays the secret of superluminal velocities, teleportation and time travel. Scalar energy transcends time and space.

Kozyrev went on to propose that space travel would be feasible using scalar energy to overcome gravity and overcome any temporal or spatial limitation. (11)

Another researcher of scalar waves, Dr. Konstantin Meyl has developed the concept of the potential vortex and explains why there are no magnetic monopoles.

Electric monopoles exist as charged particles, like the electron and proton. If space was a conducting medium, these electrical charged particles could not exist. They would be "shorted out" and immediately lose their charge. So, charged particles can only exist in insulating dielectrics, like space itself.

On the other hand, magnetic monopoles could exist if space was a conductor. But, since it is not there, are no magnetic monopoles. Magnetism requires electric current which the dielectric space does not support.

A moving charged particle creates a magnetic field circling around it. This is the magnetic vortex. If we take a coil of wire and run electric current through it, a magnetic dipole is created.

Magnetic and electric fields are related to each other by relative motion. Moving electric fields create magnetic fields and moving magnetic fields create electric fields.

So, there must be a dual to the magnetic vortex in the form of a potential vortex. This potential vortex surrounding a nonconductor would create an electric dipole.

Using the concept of the potential vortex, Meyl is able to create a unified theory of the atomic structure where the particles are potential vortexes, and unify gravity with electromagnetism.

He also experimented with Tesla's flat spiral coils and spherical antennas that he claims would transmit scalar waves. He discovered that more energy is received at the receiver than transmitted by the transmitter. He theorizes that the extra energy comes from neutrinos. These neutrinos according to Meyl are comprised of potential vortexes which become undone in the flat spiral coils and become electron charge carriers. And, this is the real secret of Tesla's magnifying transmitter. (12)

At the Earth's surface, the average neutrino flux is 66 billion neutrinos per square centimeter per second. So, there is an abundant source of energy just waiting to be tapped! This phenomenon can now be studied and examined thanks to a fully functional replica designed by Prof. Meyl.

Many of his experiments were done in his classes and observed by his students. He also sells experimental kits. Anyone can replicate his scalar wave power transmission at much lower voltages than Tesla used, and observe actual power magnification with these kits. They are sold via his website: http://www.meyl. eu/go/index.

Portals and the Cosmic Web

Now that we have been exposed to some of the secret physics, not discussed in the public universities. We are in a position to better comprehend natural portals between planets and stars otherwise known as the cosmic web.

When there is an object in space that is of sufficient mass that has spin, a natural torsion field is set up. These torsion fields link up to other objects like planets and stars, of sufficient mass and spin.

As Kozyrev and other scientists have found out these torsion fields are not limited by either space or time and allow instant communication between points that are astronomical distances apart.

The sun is a portal for space faring civilizations using a form of hyper dimensional mathematics based on sacred geometry, according to the latest testimony from secret space program whistleblower Corey Goode. He says that while serving for six years on a scientific research vessel belonging to the Solar Warden Space Program, he witnessed probes being launched into sunspots to study the sun's composition and behavior. What was discovered would shock the mainstream scientific community that believes the sun is a giant nuclear fusion generator. Instead, the sun was confirmed to be electrical in nature whose plasma outbursts acted as portals for space faring civilizations to move into and out of our solar system.

Hyperdimensional mathematics unifies, basically, all of these different scientific principles that our mainstream science has problems with. And until our mainstream science drops their theory and begins to embrace the fact that the universe is a plasmic electric universe and a torsion universe— both are true. These are the sciences that the secret space program bases their technology off of— they're not going to progress any further than we are now using this 18th and early 19th century technology. Corey Goode made the statement:

> "...NASA recently released that our sun has a basically a portal or a magnetic filament connection to every planet in our solar system, and anything that has enough mass to cause a gravitational pull or a torsion in our space-time is going to create a magnetic and gravitational relationship with the host sun. And these filaments that they are just now releasing – these filaments are the portals."

David Wilcock had in depths conversations with Henry Deacon, a government secret project insider, who said that there are ancient stargates or portal systems as well as modern portal systems. This statement is backed up by Corey Goode who states:

> "Portals are in effect an electromagnetic tube that also has a strong torsion field and would act like a traversable wormhole... There are the natural portal systems that are a part of the known universe. We call it the Cosmic Web.

And the ancient portal systems and the current era portal systems use the, or exploit, these natural Cosmic Web portal system to travel from point to point. The ancient portal systems... and there are several ancient portal systems that are left behind by several ancient groups that have been found on Earth. They vary in their sophistication. Some of them do very short point-to-point jumps to reach... Let's say you want to reach a planet or solar system that is in... They call them 'hops'. If there's, let's say 10 solar systems between where you want to go, then you may have to make three or four hops to get to that desired destination."

Corey Goode also explained that the ancient star gates were located on top of stepped pyramids many of which were discovered buried underground. He also stated that wars have been fought over these star gates which connect to many star systems. And that they have an addressing system to determine which destination you are transported to - somewhat like the science fiction series *StarGate*. But, most of these portals open up directly on top of these stepped pyramids. Only a few use a ring like in *StarGate*.

Corey also stated:

> "...before we were able to start really understanding how to use a lot of these gates, we had to have a hyper-dimensional or multi- . . . I think it was a hyper-dimensional mathematical model handed to us by an ET group."

> "... And it's happening within the torsion field of each solar system. The galaxy is a giant torsion field. All of the stars are constantly moving around the center of the galaxy and stars closer in are moving at a slightly different

speed. And the magnetic relationships are always changing. These filament relationships are always changing between each star."

So, the portal systems addressing system has to compute the proper timing and positioning to come out at the correct place.

Another problem was according to Henry Deacon, that the ancient system involves a sort of a ride or a subjective experience that can be extremely jarring. You can come out the other side at best vomiting and very disoriented and at worst completely mentally insane and irreversibly damaged by it. And he had said that people would have to study and really get advanced in their consciousness capability to be able to use these portals safely.

Corey Goode added to this:

"And also there was a chemical that was used – shots. People were given shots to help them with these effects. And this was corrected . . . they learned how to use these ancient gates more efficiently when they learned how to make the calculations better. So this became less of a problem.

But in the beginning, this kind of travel from point-to-point within our solar system was bad enough, but traveling from star to star was really not a good idea for a person. Even after we had developed it to a point where we were able to travel from planet to planet in our solar system, and negate the physical effects, it took awhile for them to get the calculations right and to fine tune traveling these ancient portal systems to travel to other star systems without these major ill effects."

Two whistle blowers have both described the traversing of portals near the Moon, Randy Cramer and Corey Goode. After leaving the Lunar Operations Command in their spacecraft the ship would hover above the moon and a transparent ceiling of the ship would allow a stunning view of the Earth hanging in the blackness of space. Then, the portal would be entered and everything went white and within minutes they would be hovering above Mars.

In addition to tapping these natural portals for interplanetary and interstellar travel. Artificial portals can be electromagnetically created. And according to Corey Goode:

> ". . . they don't have to have . . . always have to have a point-to-point . . . they don't have to have a device here and a device there. It's almost . . . They're able to pull a craft over an object and portal it onto the craft – almost like Star Trek beaming up – and it's still using the same portal-type of technology, but very advanced."

David Wilcox asked, "Can you portal yourself directly into one of these underground facilities as well as the surface of a planet? Is there any interference when you go below the surface of a planet."

Goode answered, "No, you can portal straight underground to underground of another planet of another star system."

Goode, at another time, stated that he had been deep underground:

> "I have been deep below the Earth's surface on several occasions. There are is a vast variety in the types of beings that live in the deep caverns beneath the earth. The

ecosystems down there and variety of life forms will be shocking one day when there is a "Full Disclosure Event". I have disclosed that there is not a "Hollow Earth" but a very porous "Honey Comb Earth".

There are several types of raptors and reptilians that inhabit these regions but are not friendly with each other or the human types that have networks of cities. There are also a number of ET Embassies below the surface and sea of the Earth that various ET's inhabit.

All of the reports of the Agartha Network that I saw were that they were all humans from Ancient Break Away Civilizations that had moved underground and out into bases in our Sol System and other Star Systems with a Space Program that they call the "Silver Fleet".

I have never heard of any reptilians referred to as Ancient Break Away Civilizations or apart of the Agartha Network. Some people tend to lump all cities and everything in what they believe is a "Hollow Earth" as "Agartha". This is an oversimplification and is not an accurate way to explain or depict the complexity of what is going on below the surface of the Earth and its oceans.

As far as underground Reptilians with wings being benevolent or malevolent, I have only seen reports of reptilians that reside on this planet as being of a malevolent nature by human standards (And to be avoided at all costs). It is harder to make a blanket statement about the Raptors. There are some that work very well with humans and others that will devour a human upon sight."

Corey Goode concluded on the subject of portals:

"And like I stated earlier, many of these ET races use very large craft and they portal through the Cosmic Web all over our galaxy and to other galaxies. And our galaxy is just a little bitty spot and all of the other galaxies and our local galaxy cluster and then beyond, they've already mapped out with the Hubble Telescope. They've taken pictures back of billions of years ago the energetic connections between all of these galaxies and these now are filaments that form the Cosmic Web. You've heard people say that everything in time and space is connected. That's very much the truth. It's all . . . Everything is connected and it's just . . . I mean just a short little hop skip and a jump away."

Exotic Technology

In my previous book, *Secret Science and the Secret Space Program,* I discussed science and technology that is kept secret from everyone without the proper security clearance level. Some of this technology, I had learned about outside of the traditional university courses taught at the University of California at Santa Barbara in my Electrical Engineering major and Physics minor.

There were students attending my classes that were military personnel working at the Pt. Mugu Missile Range which were being sent to school for an increase in rating. In the student lounge coffee shop we would sometimes discuss the working principles of classified technology.

Even while attending the University, I was aware of large "holes" in their course materials because I started at the University later in life and had already studied material on Thomas Townsend Brown and Nikola Tesla, which apparently were verboten (forbidden) subjects at the University.

Brown had experimentally demonstrated the connection between gravity and high voltage electric fields in the 1920's and had later patented "gravitators" and free energy devices using these gravitators.

At the University of California, in the 1980s, it was still taught that no experiment or theory that unified electromagnetism with gravitation had yet been successfully created. They also taught that the conservation of energy laws prevented the possibility of free energy.

They neglected to mention that the conservation of energy law only holds true in entirely closed systems and that in reality there are no true entirely closed systems. All systems, no matter how well sealed, are continually bathed in scalar electromagnetic energy and neutrinos, which penetrate everything.

In one Electrical Engineering class the professor stated that it was mathematically proven that longitudinal electromagnetic waves were impossible. I raised my hand and the professor pointed towards me. I pointed out that Nikola Tesla was using spherical transmitting antennas and that these type of antennas could *only* transmit longitudinal electromagnetic waves.

The professor waved me away with an expression of disgust saying "I don't want to discuss Tesla." Wow, and I idealistically had thought that universities were supposed to be places for the free expression of ideas! What I discovered was that they were places where corporate decided, science and engineering programs were pushed through with great speed and little leeway.

Interestingly, Bruce De Palma lived near Santa Barbara. He had been experimenting with a version of a free energy machine he dubbed the N machine.

He saw that a spinning conductive disk in a magnetic field will generate tremendous amperage between the axle and the outer circumference of the disk, as Faraday had originally discovered.

But, to his astonishment, he additionally discovered that the disk did not have to "cut" magnetic field lines, as was usually assumed, to generate this large current. The disk would generate current even when the magnet was attached to, and rotated with the disk!

So, now in this configuration, there was no magnetic drag on the disk. Placing a load on the generator output would not slow it down or cause the driving motor to draw more current! This reactionless outcome is totally unexplainable using classical electrodynamics.

Replications of the N machine by Paramahamsa Tewari and his Space Powered Generator were able to generate 2.38 times more power than used by the driving motor. Paramahamsa Tewari has developed theories about the electron and space power generation which are on his website here: http://www. tewari.org/

Some have theorized that the magnetic field remains stationary in space even though the magnet is rotating. But if true that would still cause a magnetic drag when current is drawn from the generator. In my own opinion, we are dealing with a magnetic vortex and torsion field effects with the N machine.

In later years, while living in the Sedona Vortex, I discovered that Sedona was also a vortex of information. I would keep running into people that worked on classified projects and free energy and exotic technology experimenters. Patrick Flanigan, author of *Pyramid Energy* and inventor of the neurophone, that allowed deaf people to hear, lived nearby.

I actually got to try a neurophone out. It was a wooden box with electronics inside and two copper tube electrodes on insulated wires coming out of the box. Apparently, It contained a recording of music inside. I would hold two electrodes of the device in my hands and could then hear music inside my head. On releasing the electrodes, the music would disappear from inside my head. It seemed pretty miraculous!

The military would later take Flanigan's neurophone and turn it into the "Voice of God" weapon. Instead of requiring electrodes, combining the neurophone concept with a microwave transmitter, they could remotely project sounds and words into people's minds.

Another fellow, who wishes to remain unnamed, had worked at Los Alamos, and after about 5 years had decided that I was harmless and not a CIA spy, became a friend.

He confided quite a story! He had worked in an underground laboratory where a recovered flying saucer was being inspected. This laboratory was inside one of the many caves beneath Los Alamos.

His job was to enter a recovered saucer, which was in a glass containment room, and inspect the interior. He had to wear a biological and chemical hazard protection suit and was accompanied by two other members likewise dressed. They were given orders not to touch anything - just go in and observe and report back afterwards.

After going through the glass room airlock, he entered the open entrance of the saucer. Inside, he observed three aliens, which seemed to be dead, which were laying prone on their backs in their pilot seats.

Their suits consisted of a skin tight, one piece suit with no zippers or other type of fastening devices. My friend became curious how the alien got in and out of their suits, which seemed an impossible feat to him.

Momentarily forgetting his orders not to touch anything, my friend touched the place where the suit ended and the alien's bare neck began.

Suddenly, the alien opened it's eyes. This panicked my friend and he immediately ran for the exit hitting the alarm button on the way out. His two comrades quickly followed him out.

He was soon taken to HQ and debriefed for hours and then told to take a shower and get some rest.

The next day, he was ordered to enter the saucer again. Wearing the same hazard suit, he returned to the scene of the previous day. Only this time, there were no aliens. Apparently, they had been removed by another team.

At this point in his story, I asked him where they took the aliens. He replied that he didn't know. Everything there was, like the CIA, on "a need to know basis" and apparently he didn't need to know.

I have personally spoken with two other people that have seen flying saucers inside of highly secured. military hangars. One at Edwards AFB, and another at an airstrip behind the Philadelphia Naval yard.

I also befriended a free energy researcher that I had met earlier at the International Tesla Society symposium at Colorado Springs in 1993 – where I had seen my first working free energy machine in the form of Joseph Newman's *Energy Machine*.

This other person had only showed a video of their device which he and his partner had developed in their lab in New Jersey. The device consisted of two counter rotating disks of copper lined with magnets and having a high voltage Tesla coil feed high voltage AC into the counter rotating disks. After the video presentation in which he claimed that free energy was being created, he stated that there were other effects of their machine that he preferred not to discuss.

Again, it took about 5 years of knowing this fellow, a time in which his partnership back in New Jersey fell apart and he divorced his pretty wife, that he finally trusted me enough to tell me the whole story of their free energy device.

For one thing, every time they ran the device, a black helicopter would appear and fly low over their laboratory and keep circling the area. This would make his partner very paranoid and eventually cause the partnership in their device to dissolve and the project to end.

The other quite interesting effect was then revealed to me. While the device was running, he would notice that a nearby metal filing cabinet started to become transparent. Out of curiosity, he would walk over to the transparent cabinet and touch it with his hand. To his further surprise – his hand passed right through the transparent metal of the cabinet!

So, here some inventors had stumbled across some pretty exotic technology in the 1990s, but were afraid to develop it further. Also, it seemed that it was putting out a field of energy that agents of the secret government knew about and could detect and send out some unmarked black helicopters to investigate and harass the inventors enough to stop their endeavor.

Thomas Bearden has written a number of books describing how woefully inadequate Classical Electrodynamics is today and demonstrated that what is passed off as Maxwell's equations are only a subset of his quaternion equations which were truncated by vectorizing them.

He has also theorized that time, like mass, is also energy. $E = TC^2$. Bearden also developed the Motionless Electrical Generator (MEG) from the Flux Switch concept of Timothy Trapp. The MEG is a free energy generator which others have replicated. It can put out much more energy than it takes to drive it.

Nassim Haramein has developed a theory along with others that demonstrates that so called empty space contains almost an infinite amount of mass/energy. Also this theory shows that the mass of a proton is much greater than previously thought and the proton is actually a mini black hole. The gravitational attraction of this greater proton mass is sufficient to overcome Coulomb repulsion between like charged protons in the atomic nucleus – thus doing away with the need for the hypothetical Strong Force!

The Russian scientist, Nikolay A. Kozyrev, experimented with torsion fields and waves. All spinning objects create torsion fields. Since magnetism is caused by electron spin, magnets also create torsion field., Planet and stars that spin also create torsion fields.

Kozyrev experimentally discovered that torsion fields travel at speeds much greater than light, perhaps instantaneously. He also discovered that they can pass through most conductive metals, except aluminum which is a good shield of torsion fields.

Most of the experimental published scientific work on torsion fields and has been done by Russian scientists.

However this does not mean that torsion field work is a Russian only science. There may be much secret work being done by other actors in this new field of scientific endeavor.

Changes in electrical resistance of various conductors is not the only effect of the influence of torsion fields. It is necessary to emphasize that torsion fields can be detected by a variety of methods. The influence of a torsion field upon a physical material results in the change of a spin state of not only this material, but alterations of the spin state of the physical vacuum. This can result in changes to a light beam's polarization angle, and change to the spin state of a substance and can result in alterations of its magnetizability, Hall's coefficient, thermal conduction, and other properties.

Changes in the spin state of an electrical conductor may result in the alteration of its electrical resistance. An elementary torsion field detector can be based upon a Wheatstone bridge. This type of detector was first utilized by N.A.Kozyrev , and later by an academician of the Russian Academy of Sciences M.M.Lavrentiev and others.

Another type of elementary torsion field detector is the torsion balance. Torsion balances were employed in experiments conducted by N.P.Myshkin at the end of the 19th century, and later were employed in the experiments of N.A.Kozyrev and others. As discovered by N.A.Kozyrev, the direction of motion of the torsion balance indicator depends upon the orientation of the torsion field.

For instance, if the torsion balances are subjected to the

influence of a "right" torsion field, and the indicator moves in one direction, then after influencing the torsion balance with a "left" torsion field, the indicator will move in the opposite direction.

Torsion fields are able to change the rate of any physical process, for instance, they significantly alter the oscillation frequency of quartz crystals. Thus this property can be employed in torsion field detectors. The effect upon the oscillation frequency of a quartz plate by torsion radiation was experimentally discovered by N.A.Kozyrev, and later was employed in various torsion detectors developed by a member of the Belarus Academy of Sciences, A.I.Veinik.

A.I.Veinik used the term "chronal detector," since he assumed a connection between the detected fields and the rate of the flow of time. He experimentally discovered that it is possible to alter the rate of any process (including the process of a radioactive decay) by subjecting that process to the influence of torsion radiation. This fact is stipulated by the ability of torsion fields to affect the inertial forces in any circulating mechanical system. This was demonstrated rigorously by G.I.Shipov.

Since the superposition of a torsion field and a gravitational field in a local area of space may result in the reduction of gravity in this area (the so called "torsion compensation of gravity"), then the influence of torsion radiation upon any physical object may result in a reduction in weight of that object.

This significant property of torsion fields was discovered in the 1950s by N.A.Kozyrev, and later, it was confirmed in the investigations conducted by A.I.Veinik, M.M.Lavrentiev, and others.

G.I.Shipov used a geometry of absolute parallelism (A_4) with an additional 6 rotational coordinates, and on the strict level it showed that the movement of any object should be described by 10 movement equations but not by 4 equations as it is in Einstein's GR. From Shipov's vacuum equations, every known fundamental physical equation (Einstein's, Young-Mills', Heisenberg's, etc.) can be deduced in completely geometrised form. G.I.Shipov showed that besides the two known long-range physical fields - electromagnetic and gravitational - there exists third long-range field possessing significantly richer properties: the torsion field.

The torsion field is an extremely unusual entity. First of all, the upper limit for the speed of torsion waves is estimated to be not less than 10^9 c, where c is the speed of light. Secondly, torsion fields are able to propagate in a region of space which is not limited by the light cone. That means that torsion fields are able to propagate not only in the future but in the past as well. Thirdly, torsion fields transmit information without transmitting energy. Fourth - torsion fields are not required to follow the superposition principle

Also inventors have developed torsion field communication devices. These devices could be used for (and perhaps already are) instantaneous communication in interplanetary and interstellar space.

So, we see that that there is much, that presently accepted science cannot properly explain and doesn't know about. As Cory Goode has said "After full disclosure, much science will have to be changed and many equations will have to be rewritten."

In *Secret Science and the Secret Space Program,* I showed experiments that showed how scalar electromagnetic waves could alter the flow rate of time, an effect originally discovered by Wilbert Smith, using a bifilar coil wound around a ferrite core, with current running in opposite directions in each of the 2 windings. He called his discovery a "Tensor Coil" because of its ability to change the time flow rate.

The scalar waves generated by this tensor coil could also pass through Faraday cages and conductors like sea water. Another experimenter, T. Townsend Brown, developed gravity wave detectors and transmitters that showed similar effects. (ie. the ability to be transmitted through conductive media.) Thomas Bearden says that scalar waves are also electro gravitational waves.

Thomas Bearden says that scalar waves only exist in the time domain because they have no vectors in 3 space. Since these waves only wave in the time domain, they don't travel through space. Since they don't travel through space, but along the time axis where each point in space is everywhere at the same instant of time, they arrive instantaneously at their destination, violating Einstein's theory of relativity in several ways.

I also mentioned Russian experiments by Dr. Vadim Chernobrov with altering the flow of time using scalar electromagnetic waves that was tested on human subjects with subjectively interesting results.

Another scientist from Belarus, Albert Venik, followed up on the work of Dr. Vadim Chernobrov and used this scalar electromagnetic ability to slow or speed up time in an interesting experiment. He took a spinning flywheel and slowed time on one half and speeded it up on the other half. The side of the flywheel with the higher time rate developed greater centrifugal force than the half with the lower time rate. Thus, he had created a machine that could generate a propulsive force from within itself - in violation of Newtonian physics!

This sounds like pretty exotic technology already. In my previous book, *Secret Science and the Secret Space Program*, I went in detail how this technology worked, so I only touched on the subject here.

But, when we listen to the testimony of the men, like Corey Goode and Randy Cramer who worked in these secret space programs, we learn that even more exotic technology has been hidden from the people of Earth.

For example, Technology that can teleport people from Earth to Mars in jump gates. Andrew Basiago was one of the first to come forward and speak of this jump room technology which he himself has experienced both on the planet between New Jersey and New Mexico, when he was a child in the 1970s, and between El Segundo, California and the Mars Base in the 1980s. Others have also come forward and revealed that they were in the same program as Andrew Basiago, in the 1980s and had also experienced this jump room technology between Earth and Mars.

As another example, Technology that can allow spacecraft to traverse interplanetary and interstellar portals in minutes rather than days, years or decades. Randy Cramer and Corey Goode have both experienced this technology first hand. They both stated that the trip between the Lunar Operations Command on the Moon and the Mars Colony headquarters on Mars only took around a quarter of an hour.

Apparently there are wormhole like natural portals connecting the planets and the stars and even the galaxies, that can be used to transverse great distance between the planets and between the stars in minutes instead of days and years. Apparently, the extraterrestrials gave the secret space programs the knowledge on how to navigate and traverse these portals without hazard.

Another example, is technology that allows time travel. This technology is related to the jump room or teleportation technology which allows almost instantaneous travel anywhere in the time space continuum. Nikola Tesla was the first to experiment with this phenomenon, which was upgraded after the Philadelphia Experiment (Project Rainbow) and the bugs worked out.

Tesla was using rotating magnetic fields which was also generating torsion or scalar waves. As the Russian scientist Kozyrev has shown, these torsion waves are related to time. The Philadelphia experiment was also using tremendous current in their rotating magnetic fields which created a tremendous diverging Magnetic Vector Potential sufficient to create the portal condition, $k\nabla \cdot \mathbf{A} < - C^2/2$, as explained in the chapter, Scalar Physics, in this book.

Alfred Bielek, a survivor of the Philadelphia experiment and the Montauk project, explains that a zero time reference generator was needed to be developed to overcome the problems with crew members becoming embedded in the steel of the ship when the ship returned to normal space time. In the ensuing Phoenix Project, all the bugs were worked out and practical space and time teleportation became a highly classified technology.

In the 1970s, Andrew Basiago was a part of a young "Cronaut" team being trained to both teleport in space and time after it was discovered that children could mentally handle time travel better than adults. At one time, Basiago was teleported in time, back to President Lincoln's Gettysburg speech. Basiago has even produced a black and white photograph of that speech which shows his presence at that historic event.

Another version of this time travel technology is the Chronovisor, which allows a person to see into the future or past without actually traveling there. Basiago claims that this Chronovisor showed some elite persons in our government, future events like George W. Bush being elected President, or the 9/11 attacks decades before they happened.

Imagine how this technology would allow the ultimate inside traders to make tremendous profits in the stock and commodities markets.

Corey Goode also spoke in interviews with David Wilcox, which were recorded at GaiamTV, about technology that could change the phase relationships between a person and their surroundings. This would allow that person to literally walk through walls!

He also spoke of "gravity plates" that would generate artificial gravity so that people and loose objects would not float around in the space ships while in gravity free space.

In the Chapter, Scalar Physics, in this book, the electromagnetic theory and equations are presented that would allow the engineering of similar gravity or antigravity plates.

Another example are technologies that can erase selective portions of a person's memory. This technology is an offshoot of the CIA's Project MKUltra, which the CIA got from the Germans after World War II and developed further. This technology is part of Radio Hypnotic Intra Cerebral Control – Electronic Dissolution of Memory (RHIC-EDOM) technology. More detailed information on this technology is in *The Secret History of the New World Order* in Chapter 7, Propaganda and Mind Control.

Also is the medical technology that amazingly can grow new limbs, or patch up bodies blown up in battle.

According to Randy Cramer, The Secret Space Programs use all of this technology, particularly in the Mars Defense Force. As a super soldier, he himself was blown up and his life saved a number of times with this technology.

Corey Goode also spoke of Star Trek like "replicator" technology used on his spaceship, which they actually called "printers." These "printers" could replicate many things like a hot roast beef dinner, which tasted pretty good according to Corey Goode.

Imagine how it will change the world when this technology is openly known about, understood and widely used. Then, we will truly have a Star Trek civilization.

After Disclosure

The U.S. Government has lied to the people of the United States since the end of World War II. There may have been some Government lies before then, but the lying got really serious after the war. This time frame corresponds with the time the Nazi International was infiltrating our government.

We have been lied to about the extraterrestrial presence on our planet, about the Cold War, about why we were in Vietnam, and every war since. We have been lied to about the horrible crimes against humanity committed by our government leaders - most of which are war crimes. We have been lied to about the War on Drugs, the War on terror, geoengineering, and the 9/11 attacks.

Where outright lies weren't fed the U.S. public - secrecy did the job. Secrets like the Secret Space Programs, the Deep Underground Military Bases (DUMB), the secret Muslim Brotherhood covert terrorist army of the CIA, the secret CIA illegal narcotics trafficking to fund "black Projects". The list goes on and on.

There are some that would like to make the United States great again. I am in full agreement with that concept. However our country will never be truly great if it is living a lie. The first step in making our country great again will have to be the full disclosure of all the skeletons in the U.S. closet.

This will reveal the crimes committed by government leaders, military leaders, international corporations – especially banking corporations, and intelligence agencies. When full disclosure finally happens – and I am confident that time will be soon - the public shock will be tremendous. There will be public demands for indictment and trials of the traitorous criminals involved. And, that will be a necessary first step in cleansing our nation of the criminal filth that has been running it for far too long.

In my *CIA: Crime Incorporated of America*, I only touched the tip of the iceberg of criminality of just one government agency from my own researches. Any government agency that hides under the cloak of "National Security" has way too many means of committing crimes and hiding them - they *all* need to be thoroughly investigated.

When full disclosure of the slave labor being used in the ICC run colonies on Mars and other places, the CEOs and managers of these corporations will also have to face indictment for crimes against humanity.

Much of this present disclosure is coming from Corey Goode. He was asked if he wasn't placing his life and that of his family in danger by his role in the disclosure process. This is what he had to say:

> "I had not planned on coming out into the open. This decision was made for me by the executive decisions of some researchers who have lost their way ethically. It

does appear that it was apart of the overall plan for me to come out eventually. I would preferred to have come out as my full identity under my own terms after I had taken steps to put in further security measures.

I and my family have not gone through this process with zero interference or death threats.

I had someone leave a note and a bullet in my mailbox at my home. I have been directly threatened by different "cabal" associated groups and individuals and have had a coordinated campaign to discredit me through compromised researchers, bloggers and triggered individuals in the "Truther Field".

This field is completely infiltrated and has been since it began in the 1950's. It has been manipulated and controlled by operatives since its inception.

Those in it who believe this manipulation couldn't happen to them are speaking out of Ego and are most likely already targets themselves. There have been some researchers in the field that have become psychologically compromised and could not resist writing themselves into the narrative of what they were researching. This has all been crafted by design by those who have infiltrated and manipulated (handled) people in this field. This is not me speaking out of spite against some who have targeted me but is just an uncomfortable truth.

Those who say I have not had death threats or been threatened are misinformed. The opposite is true.

We even had a couple of incidents where my family has

been terrorized. Being that there are such vicious trolls and unethical bloggers/researchers out there who are ready to pounce on anything I say... I make sure to keep information that includes my family out of the public eye. These people will then make a target of them or accuse me of making up stories of their very real experiences.

Make no mistake that we are putting our lives and reputations in extreme jeopardy and have already payed some serious costs to date. It has caused fear and anxiety in my family because they have not felt safe in their own home at times.

If not for the SSP Alliance and the Sphere Being Alliance wishing me to come forward with these disclosures it would not be happening. The Secret Earth Governments and their Syndicates have kept a very tight reign on the Ufology/Truther Field from the beginning. If not for these opposition groups this status quo would be in effect."

So, being a source of secret information is not without it's costs and was not an easy decision for Corey Goode to make. But, soon there will be other men and women who come forward to tell the people what is really going on out in space, as others already have.

I am of the opinion that the public shock will be much greater over the revelation of government criminality, than that of the extraterrestrial presence or the secret space programs. In any case, Full disclosure will bring many changes to the planet. These changes will vastly improve the living conditions for *all* of humankind - not just the financial elite.

After 2008, when the government of Iceland arrested the

bankers that caused the financial collapse in their country, instead of bailing out the banks with taxpayer money, as the U.S. government did, their economy improved considerably.

Iceland's economy is now better than that of most of the E.U. countries which caved in to the European Central Bank (ECB) and International Monetary Fund (IMF) demands for austerity.

The truth is that the IMF has been practicing predatory economics since it's creation in Bretton Woods in 1943. I won't go into detail here, as anyone can read *The Confessions of an Economic Hit Man* by John Perkins to see what predatory tactics are used to fleece developing countries of their wealth by the IMF and other corporations.

Africa has been economically raped considerably, first by the European colonists and later by the multinational companies. Latin America, Indonesia, and the Philippines also face similar problems. These injustices will also be addressed along with criminal trials of corrupt government leaders that cater to multinational corporations (usually via bribery) rather than their own people.

Countries that depend on exporting oil and natural gas will have to diversify into other means of trade and commerce. The burning of oil and natural gas for fuel will soon become obsolete and the demand for it will dramatically decrease in the decades that follow full disclosure.

Petrochemicals will be used only for the creation of plastics and other useful products and will not be wastefully burned as fuel, while polluting the atmosphere.

Free energy technology will be non polluting and free of fuel costs. The mass production of free energy devices will make them affordable and they will be incorporated in every home and means of transportation.

This will do away with the ugly and vulnerable to EMP attack, power grid that most are now overly dependent on. People will no longer pay expensive power bills or pay for fuel. This alone will stimulate the economy as people will have more to spend on other goods. This will also bring about more travel because zero fuel costs will bring travel expenses down.

Eventually people will have their own flying saucers to travel around in just as they have cars now. These saucers will have unlimited range since they use the energy that is everywhere available in space.

A true one world government that caters to the good of all, and not just the ultra rich, will come about. People will be able to travel anywhere on the planet with their own personal flying saucers and national borders will be unenforceable and non existent. Passports and IDs will also be a thing of the past, as people will be known by their vibrations and thoughts because of the greater intuitive and telepathic powers awakening in us because of the higher vibrational rate the Earth and it's inhabitants are experiencing.

Money will no longer be wasted on destructive war. People who wish for war will no longer be living on this planet as it's vibratory rate increases. These people of lower spiritual vibration will die and will reincarnate on planets of lower vibrations to work out their karma via more pain and suffering with others of like kind. The Earth is ascending. This is the opportunity of the ages. It is up to us to ascend with our beloved planet.

When the surface people learn to abide and actually live by the teachings of the spiritual teachers sent them, like Jesus Christ, Zoroaster, Buddha and others, the Agarthan civilization beneath the surface will rejoin their brothers and sisters on the surface of the planet as they have already promised to do. This will be the beginning of Paradise on Earth.(13)

The public ownership of replicator machines, which Corey Goode says are presently being used on craft of the Secret Space Program will also cause dramatic changes.

Money and Jobs will both lose importance in the public viewpoint. Replicators that can replicate virtually anything will mean that anyone could create as much money as they wanted in their home replicator machine.

"The love of money is the root of all evil." This famous saying has much truth. People love money for the things that it can buy – not for the paper itself. So with your replicator you can bypass money and just make what you want directly. This device alone will probably revolutionize the whole economic system.

Robotic production will mean that human labor will be greatly lessened. We may have to go to a 3 day work week with rotating shifts to double the number of humans employed.

People will work at things because they want to - not because they are economically forced to, as in the present economic system. All people need to feel worthwhile and require some form of meaningful work for their own sense of well being.

This will also mean that most paper shuffling work and "financial products" will also become obsolete as people lose interest in things like insurance, financial derivatives, mortgage backed securities, credit default swaps Etc.

The stock market and trading of commodities will also lose its glamor as people will lose interest in the game and start to work at being truly happy in place of the abstract goal of becoming happy at some future time with their hypothetical trading gains.

And as greater spiritual evolvement and wisdom overcomes mankind, people will derive more pleasure from helping others than themselves.

They will not be like the idiots and spiritual dwarfs presently trying to own and control the planet and who are such terrible role models for the rest of us. Ever notice how the corporate controlled media glorifies rich people and tries to plant the desire to become like them? Usually, when one looks closer into these famous lives, they actually find a great deal of unhappiness.

First of all, money does not automatically bring happiness. But, having a good heart does. "What profiteth a man who gains the whole world and loses his own soul in the process?" When you die, you have to leave all your material possessions behind anyway. But, you do have to take your soul with you. And your soul will determine what lays in store for you in the next world!

So what kind of madness possesses people that always want more and more even though they already have thousands of times more than everyone else? Pray that these sickos become enlightened before it is too late for their souls. I sure would not want to be in their shoes.

Not that all rich people are bad. Many of them got their wealth by creating beneficial things for humankind and are actually motivated to be of service to others. They actually create jobs and are helping to stimulate the economy and make life better for most of us.

And then there are the ones that are pure predators and only motivated by selfish greed. It doesn't take long to figure out the difference between the two types, by people with discernment who bother to really inform themselves.

We cannot change other people. It would be wrong to try. That would be a form of fascism that is practiced by insecure people. We can however change ourselves - which is our true responsibility. We can try to become more spiritually evolved by being of service to others more than to ourselves. And as we do, we will bring the world one step closer to the future paradise the world is destined to become.

In any case, full disclosure, along with our own actions, will dramatically change the world that we presently live in. Believe it and live it!

End.

Other Books by this Author:

1. The Adventurer http://www.amazon.com/Adventurer-Auto-biography-Herbert-Grove-Dorsey-ebook/dp/B00547RVQK

2. The Secret History of the New World Order http://www.amazon.com/Secret-History-New-World-Order-ebook/dp/B00LAECCJM

3. Secret Science and the Secret Space Program http://www.amazon.com/Secret-Science-Space-Program-ebook/dp/B00PSS3RE0/

4. CIA: Crime Incorporated of America http://www.amazon.com/CIA-Incorporated-Herbert-Dorsey-III-ebook/dp/B013RJEJ20

Bibliography

1. *Dark Star* by Henry Stevens, Part Three: Zusammenfassung

2. *Secret Science and the Secret Space Program* by Herbert G. Dorsey III

3. This information is from various sources including *Cosmic Disclosure season 2 episode 1* of David Wilcox and Corey Goode interviews on GaiamTV.

4. *The Nazi International* by Joseph P. Farrell

5. *Britain's Secret War in Antarctica* Nexus Magazine Volume 12, Number 5

6. *Hitler and the Secret Alliance* by Michael Ivinheim

7. *Sknorzeny, Hitler's Commando* by Glenn B. Infield

8. *Exposing U.S. Government Policies on Extraterrestrial Life: The Challenge of Exopolitics* by Michael E. Salla, Ph.D.

9. http://spherebeingalliance.com/blog/joint-ssp-sphere-alliance-icc-leadership-conference-tour-of-mars-colony-on-6-20.html

10. http://www.scalarphysics.com

11. http://tompaladinoscalarenergy.com/nikolai-a-kozyrev-scalar-energy-astrophysicist/

12. *Scalar Waves* by Konstantin Meyl.

13. *The Kingdom of Agartha* by Marquis Alexandre Saint Yves D'Alveydre and *Telos (Vol. 1 –3)* by Aurelia Louise Jones.

CPSIA information can be obtained at www.ICGtesting.com
Printed in the USA
BVOW02s1543300716

457020BV00002B/133/P

9 781478 768838